W9-CAM-108

The Oxford A to Z of Word Games

The Oxford A to Z of Word Games

Tony Augarde

Oxford New York

OXFORD UNIVERSITY PRESS

Oxford University Press, Walton Street, Oxford OX2 6DP

Oxford New York
Athens Auckland Bangkok Bombay
Calcutta Cape Town Dar es Salaam Delhi
Florence Hong Kong Istanbul Karachi
Kuala Lumpur Madras Madrid Melbourne
Mexico City Nairobi Paris Singapore
Taipei Tokyo Toronto
and associated companies in
Berlin Ibadan

Oxford is a trade mark of Oxford University Press

British Library Cataloguing in Publication Data
Data available

Library of Congress Cataloging in Publication Data
Data available
ISBN 0-19-866178-9

10 9 8 7 6 5 4 3 2

Typeset by Pentacor plc
Printed in Great Britain
on acid-free paper by
Biddles Ltd
Guildford and King's Lynn

TO ANNA, KATE, AND CHAS

Contents

Preface

This dictionary describes more than 250 games that can be played with words. Language is a superbly rich resource: not only for writing and speaking but also for playing games. One can play with the meanings of words, the sounds of words, or the spellings of words. One can play with the letters of words: shuffling them around or arranging them in particular sequences. There are games based on the alphabet, and games that arrange words in patterns or interlocking grids.

One of the most attractive features of word games is that most of them are very simple and require little or no equipment. Instead of spending a large amount of money on a computer game or boxed game, one needs only a pencil and paper—or simply voices—for hours of innocent enjoyment. Some word games are competitive but in many games everyone is a winner: gaining pleasure from the mental stimulation, the challenge, the social interaction, and the sheer fun that such pastimes provide.

This book is a kind of companion to *The Oxford Guide to Word Games*, in which I described the history and social background of the best-known games. This dictionary is intended to be more comprehensive and more utilitarian: describing how to play all the word games worthy of inclusion. It is difficult to draw the line between word games and other games. Some games (like *Guggenheim*) use words but are also concerned with the things that words represent. It is also difficult to distinguish between word games and various forms of wordplay: the latter are generally included if they have the character of a game or form the basis for a game.

The main entries classify each game according to the number of players, the type of game, how it is played, and what equipment (if any) is required. The basic object of each game is summarized in one sentence, and then the procedure is described in more detail. A special feature of this dictionary is that it gives examples of actual play for many of the games, using the same group of four players. These examples include hints on strategy and insights into the best—and worst—methods of play. The dictionary is deliberately designed for browsing as well as reference.

Tony Augarde

How to use this Book

Order The games are listed in strict alphabetical order, so that—for example—*Alphabetical Dinner* precedes *Alphabet Race*, and *Word-Building* precedes *Word Squares*. Cross-references are given for every game that is described elsewhere in the dictionary. The thematic index will help readers to find games of particular types or for specified numbers of players.

Playing the games Before starting to play most word games, players should choose a dictionary for arbitration in the event of disagreements about the acceptability of particular words. In most games, players are not allowed to use proper names, foreign words, and abbreviations. Some players also prefer to disallow plurals and inflections, but the choice of a good dictionary should avoid disputes. *The Concise Oxford Dictionary* (8th edition) has been used as the main source of reference for the examples in this book. Remember that you can agree on your own rules for many games—but this is always best done before the game begins. Unless otherwise stated, the winner of each game is the person who scores the most points or the one who is left after everyone else is 'out'. However, winning these games is much less important than deriving pleasure from playing them.

Types of games Most of the games are either written (and require pencils and paper) or spoken (and need no equipment). Many games can be written or spoken. This information is given at the start of each entry, together with an indication of what type of game it is. Most of these categories are self-explanatory but the following categories may need clarification.

- *Active* games involve physical activity or movement.
- *Challenge* describes games which present players with a challenge: something to solve or achieve.
- *Cumulative* games are those which build up gradually through the contributions of all the players.
- *Grid* games arrange words so that they interlock or form a pattern.
- *Word-building* games involve building up words, usually one letter at a time.
- *Word-finding* games involve finding words that are hidden in some way, or thinking of words to fit stated categories.

Index: Number of Players

Games for only two players

Alphacross	Get the Message	Quizl
Arrow of Letters	Jarnac	Ragaman
Black Squares	Jotto	Sinko
Chesterfield	Lucky Ladders	Wild Crash
Convergence	Oilers	Word Battleships
Crash	Pi	Wordrum
Double Jeopardy	Questions	Word Squares (2)

Games for two or more players

Abbreviations	Build-Up	Fusions
Acrostics (2)	Catchword	Game
Add a Letter	Categories	Geography
Add-a-Word	Centurion	Ghosts
Adverbs	Charades	Good News, Bad
Alphabet Dinner	Clue Words	News
Alphabetical	Coffeepot	Grandmother's
Adjectives	Combinations (1)	Trunk
Anaghost	Comparisons	Group Limericks
Anagrams (1)	Concealments	Guggenheim
Anagrams (2)	Connections	Hangman
Anaquote	Consequences	Headlines
Antonyms	Countdown	Heteronyms
Antonyms and	Crambo	Hobbies
Synonyms	Crossword Game	Homonyms
Associations	Definitions	Homophones
Avoid 'em	Dictionary Game	Hypochondriac
A Was an Apple	Digrams	I Love My Love
Pie	Dumb Crambo	Inflation
Awful Authors	Encyclopaedia	Initial Letters (1)
Backenforth	Fictannica	Initial Proverbs
Backward Spelling	Endings	Initial Sentences
Bee	Fill-Ins	I Packed My Bag
Blockbusters	Find the Word	I Spy
Botticelli	Follow On	I Went to Market

Journalism
Kan-U-Go
Keyword
Kolodny's Game
Last Word (1)
Letter Auction
Lexicon
Lynx
Minister's Cat
Mischmasch
Missing Letters
Missing Vowels
Monosyllables
Nerbs
Numwords
Odd One Out
One Awkward
  Albatross
Outburst!
Paring Pairs

Pass It On
Pictures
Portmanteau Words
Printers' Errors
Proverbial Answers
Proverbs
Reversals
Rhyme in Time
Rhyming
  Consequences
Rigmarole
Sausages
Scrabble
Scramble
Side by Side
Spelling Bee
Stairway
Starters
Stepping Stones
Stinky Pinky

Superghosts
Syllables
Synonyms
Taboo
Tennis, Elbow, Foot
Tongue-Twisters
Tops and Tails
Traveller's Alphabet
Triple Meanings
Twenty Questions
Uncrash
What is My Thought
  Like?
What is the
  Question?
Who Am I?
Word-Making
Word Ping-Pong
Yes and No (1)

Games for three or more players

Ad Lib
Aesop's Mission
Authors
Buzz
Chinese Whispers
Foreheads
How, When, and
  Where

Jumbled Proverbs
Just a Minute
Last Word (2)
My Name is Mary
Name Game
Punchlines
Questions and
  Answers

Railway Carriage
  Game
Simon Says
Spelling Round
Throwing Light

Games for four or more players

A What?
Birds, Beasts, Fishes,
  or Flowers
Clumps

Cross-Questions
Earth, Air, Water
Password
Shouting Proverbs

Talkabout
Word Parade

Index: Types of Game

Active

A What?
Birds, Beasts, Fishes,
   or Flowers
Charades
Game
Simon Says
Word Parade

Alphabetical

Alphabent
Alphabet
Alphabet Dinner
A Was an Apple Pie
Centurion
Hypochondriac
I Love My Love
Inflation
I Packed My Bag
I Went to Market
Minister's Cat
One Awkward
   Albatross
Traveller's Alphabet

Anagrams

Add a Letter
Alphabetical
   Adjectives
Anaghost
Anagrams (1)
Anagrams (2)
Antigrams
Beheadments
Countdown
Dumb Crambo
Huntergrams
Isosceles Words
Jumble
Last Word (2)
Logogriphs
Transadditions
Transdeletions

Challenge

A A
Ad Lib
Alpha
Alphabent
Alphabet
Autantonyms
Authors
Automynorcagrams
Avoid 'em
Bouts-rimés
Bullets
Buzz
Centos
Chronograms
Deflation
Doublets
Earth, Air, Water
Endings
Equivoque
Famous Last Words
Follow On
Head to Tail
Hobbies
Initial Letters (1)

Cumulative

Grid Games

Addiction
Alphacross
Black Squares
Boggle
Chesterfield
Cross Wits
Crossword
Crossword Game
Double-crostic

Get the Message
Jaxsquare
Knight's Tour
Last Word (1)
Lynx
Pi
Quizl
Ragaman
Scrabble

Scramble
Sinko
Trackword
Word Battleships
Word Search
Word Squares (1)
Word Squares (2)

Guessing

Abbreviations
Adverbs
Aesop's Mission
Alternade
Birds, Beasts, Fishes,
  or Flowers
Blockbusters
Botticelli
Charades
Chronograms
Clue Words
Clumps
Coffeepot
Combinations (1)
Comparisons
Connections
Convergence
Crambo
Crash
Definitions
Dictionary Game
Double Jeopardy
Dumb Crambo

Encyclopaedia
  Fictannica
Enigma
Fill-Ins
Foreheads
Fusions
Game
Get the Message
Hangman
Heteronyms
Homonyms
Homophones
How, When, and
  Where
Initial Proverbs
I Spy
Jotto
Jumbled Proverbs
Knock-Knock
Kolodny's Game
Lucky Ladders
Metagram
Missing Letters

Missing Vowels
Name Game
Odd One Out
Paring Pairs
Password
Pictures
Proverbial Answers
Proverbs
Quizl
Rebus
Reversals
Riddle-Me-Ree
Riddles
Shouting Proverbs
Shrink Words
Side By Side
Stinky Pinky
Talkabout
Throwing Light
Tops and Tails
Triple Meanings
Twenty Questions

Spelling

Backenforth	Backward Spelling Bee	Spelling Bee
		Spelling Round

Word-Building (*see also* Grid Games)

Addiction	Digrams	Lexicon
Arrow of Letters	Jarnac	My Word (2)
Boggle	Kan-U-Go	Syllables
Build-Up	Last Word (2)	Trigrams
Catchword	Letter Auction	Word-Making (1)

Word-Finding

Acrostics (1)	Guggenheim	Stairway
Acrostics (2)	Hidden Words (1)	Synonym Chains
Alpha	Insertion-Deletion Network	Synonyms
Anaquote		Telegrams
Antonyms	Kangaroo Words	Tom Swifties
Antonyms and Synonyms	Keyword	Tops and Tails
	Knight's Tour	Trackword
Associations	Letterbank	Traveller's Alphabet
Categories	Metagram	Triple Acrostics
Concealments	Mischmasch	Word Search
Double Acrostics	Numwords	Words Within Words
Fill-Ins	Outburst!	
Find the Word	Proverb Delve	
Geography	Reversals	

Wordplay (*see also* Punning)

Anguish Languish	Semantic Poetry	Word Substitution
Malapropisms	Spoonerisms	
Oxymorons	Word Divisions	

A

A A

⓸ Any number of players	✎ Played by writing
? Challenge game	✎ Played with pencils and paper

Object To write a piece consisting of words that all start with the same letter.

Procedure Each player is given a piece of paper and is told to write one letter of the alphabet at the top of it. Players are then told that they have to write a story, news report, poem, or biography—using only words that begin with the letter they have chosen. This game can also be used as a challenge for one person to tackle, or more than one person to share.

Example An Arab artist achieved amazing acts as an architect arranging all areas around an arbutus . . .

Abbreviations

⓸ Two or more players	✎ Played by writing and speaking
? Guessing game	✎ Played with pencils and paper

Object To guess a phrase or saying from its initials.

Procedure One player writes down on a piece of paper the first letters of the words of a well-known phrase or saying, such as a proverb, quotation, book title, etc. All the other players are told what type of phrase it is, and they try to guess the phrase from its initials. The first person to guess the phrase chooses a new phrase for the other players to solve.

Example Tony writes down the initials HWHIL and tells the other players that these are the initial letters of a five-word proverb. They rack their brains and eventually Kate guesses correctly that the proverb is 'He who hesitates is lost.' Kate then writes down ITFEWBFFFM and tells the others that it is a quotation by a twentieth-century artist. The other players have difficulty guessing it, so Kate supplies the clue that the quotation is by Andy Warhol. Anna immediately recognizes the saying as: 'In the future, everyone will be famous for fifteen minutes.'

Acromania *see* Initial Sentences.

Acronymia *see* Initial Sentences.

Acronyms

⓿②③	Any number of players	✍	Played by writing or speaking
?	Letters game	✎	Played with pencils and paper, or with no equipment

Object To devise an appropriate phrase consisting of words beginning with the letters of a chosen word or phrase.

Procedure Players choose—or are given—a word, name, or phrase. They try to make a sentence or phrase using words which start with the letters of the chosen word or phrase. The result should describe or comment on the chosen word(s), preferably with humour.

Examples

SNAIL: slimy nocturnal animal invading lettuces.

ALITALIA: always late in take-off; always late in arriving.

CRICKET: crowd rudely ignores contest, keeps eating tea.

AMADEUS: anguished Mozart's accelerated death engineered under Salieri.

STAR WARS: space, time, and relativity, with a ridiculous script.

Acrosticals *see* Acrostics (2).

Acrostic Contest *see* Acrostics (2).

Acrostics (1)

② ③ Any number of players	✍ Played by writing
？ Word-finding game	✎ Played with a previously prepared puzzle, plus pencils and paper

Object To solve clues leading to words whose first letters spell another word.

Procedure An acrostic is a puzzle in which clues are given to words whose first letters spell out a mystery word or phrase. The clues are sometimes given in rhyme.

Example These six clues lead to six words, whose initial letters spell a famous city:

1. It's beside the road.
2. This involves a load.
3. This is pleasant for you.
4. And a song for two.
5. A European river.
6. Makes them all a-quiver.

Compare Double Acrostics; Triple Acrostics.

Solution

LAY-BY

ONEROUS

NICE

DUET

ODER

NERVES

Acrostics (2)

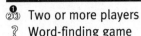 Two or more players
? Word-finding game

✍ Played by writing
✎ Played with pencils and paper

Object To think of words that start and end with letters determined by one word written forwards and backwards.

Procedure The first player chooses a word of at least four letters—usually six or more—which each player writes in a column down the left-hand side of their paper. They then write the same word *in reverse* down the right-hand side of the paper. Players have to fill in words which start and end with the resulting letters. The winner is either the player who fills in all the spaces first or the player who uses the longest word or words within a set time (usually five minutes).

Example Chas suggests *English* as the word. He fills in the words like this, later adding the scores for each letter used:

E   a   c   H = 4
N   a   m   e s = 5
G   o   b   I = 0
L   a t e r a L = 7
I   n   K i n   G = 6
S   o l u t i o N   8
H a p p e n s t a n cE = 12
                          ‾‾‾‾‾
                          42
                          ‾‾‾‾‾

4

Kate fills in the spaces thus:

E a r t H = 5
N i c k n a m e S = 9
G u a r a n I = 7
L i t e r a L = 7
I n d u l g i n G = 9
S e v e N = 5
H u g E = 4
          46

Depending on the method of scoring, Chas would win for having the longest word or Kate would win for scoring the highest total. The word *Gobi* was disqualified because it is not in the *Concise Oxford Dictionary* (chosen as the reference source before the game started); but *Guarani* was acceptable because it *is* in this dictionary.

Also called Acrosticals; Acrostic Contest; Dictionary.

Acting Rhymes *see* Crambo.

Acting with Adverbs *see* Adverbs.

Add a Letter

| ⓐⓑ Two or more players | ✍ Played by speaking |
| ? Anagrams game | ✎ No equipment needed |

Object To solve anagrams consisting of words with one added letter.

Procedure One player gives another player a word plus an extra letter. That player has to make a new word by rearranging these letters. In some versions of the game, a clue to the new word is given by the first player.

Example

KATE: Can you add an M to *beers*, rearrange the letters, and make a
   word for ashes?

CHAS: *Embers.* Can you add an O to *hard* to make treasure?

ANNA: *Hoard.* Can you add an A to *pared* to make a procession?

TONY: *Parade.* Can you add an A to *gleans* to make an Italian food?

KATE: Er . . . oh, yes! *Lasagne!*

Compare Transdeletions.

Add a Word

⓪②③ Two or more players	✍ Played by speaking
❔ Cumulative game	✎ No equipment needed

Object To build up a story with each player contributing one word at a
time.

Procedure The first player begins by saying one word; the second
player adds another word; and so on, gradually building up a story.

Example

KATE: One

CHAS: Day

ANNA: Kate

TONY: Saw

KATE: A

CHAS: Hippopotamus

ANNA: Walking

TONY: Down

KATE: The

CHAS: High

ANNA: Street.

TONY: 'Hallo'

KATE: Said

CHAS: Kate . . . (*And so on.*)

Compare Consequences; Pass It On; Rigmarole.

Addiction

👥 Any number of players
❓ Grid game; word-building game

📖 Played with a boxed set of a tray and 13 lettered cubes

Object To make words in a grid from letters that appear at random.

Procedure Players place one of the lettered cubes on the top part of the tray, and roll it down the 'steps' into a slot, revealing one letter on top of the cube. The player then places the cube, with that letter uppermost, somewhere in the grid of five-by-five squares. The players continue in turn, rolling one of the cubes down the steps and placing the revealed letter in the grid, trying to build up words. Anyone who makes a word will score the number of points which are printed on the cubes beside each letter (A scores two, B scores four, etc.).

Compare Crossword Game.

Background A boxed game copyright Waddingtons Games Limited.

Ad Lib

👥 Three or more players
❓ Challenge game

📖 Played by speaking
✎ Played with a clock or other timer

Object To keep a speech going continuously from one player to another.

Procedure This is a communal form of the game *Just a Minute*. One player becomes umpire and chooses a subject which the second player has to speak about for half a minute, without hesitation, repetition, or outlandish deviation. After 30 seconds, the third player has to continue the talk, and so on round the players. Anyone who is guilty of hesitation, repetition, or deviation is out of the game. The winner is the last player who is left: this player becomes the umpire for the next round.

Compare Just a Minute.

Adverbial Puns *see* Tom Swifties.

Adverbs

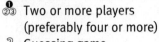

👥 Two or more players (preferably four or more)	✍ Played by miming
❓ Guessing game	✎ No equipment needed

Object To guess an adverb from the manner in which people carry out a particular action.

Procedure The people present divide into two teams. One team leaves the room and decides on a particular adverb. When this team returns to the room, members of the other team ask them in turn to carry out a particular action 'in the manner of the word'. The word has to be guessed from the way in which the actions are performed. The game can also be played with only one player going outside and returning to guess the adverb. Alternatively, the players in the room can choose the adverb, and the players outside return to ask the others a question which has to be answered 'in the manner of the word'.

Example Chas and Tony go out of the room and choose the word *painfully*. When they return, Kate asks them to act out 'making breakfast, in the manner of the word'. Tony pours out some cornflakes while apparently racked with pain, and Chas gets his fingers stuck in an imaginary toaster. Anna and Kate guess *agonisingly* and *rheumatically* but fail to guess *painfully*. Anna asks them to act out 'singing, in the manner of the word'. Chas mimes singing, while Tony holds his hands over his ears. Anna and Kate guess *painfully*, and it is now their turn to go out of the room and choose an adverb for Chas and Tony to guess.

Also called Acting with Adverbs; In the Manner of the Word.

Compare Proverbs.

Aesop's Mission

- ∰ Three or more players
- ? Guessing game
- ✍ Played by speaking
- ✎ No equipment needed

Object To guess a letter of the alphabet which is 'taboo'.

Procedure One player is chosen as 'Aesop', who has to question the other players. Aesop mentally chooses one letter of the alphabet which is 'taboo' and then asks the other players questions in turn. The players have to give an answer (usually in one word): if their answer includes the forbidden letter, they lose one 'life'. Anyone who loses three 'lives' is out of the game. Any player who guesses the forbidden letter becomes 'Aesop' for the next round of the game.

Example

TONY (*as 'Aesop', having privately chosen the letter M as taboo*): What time is it?

ANNA: Eight.

TONY: How old are you?

KATE: Twelve.

TONY (*getting sneaky*): Where is Valletta?

CHAS: Malta.

TONY: You lose one life. Which white jewel is related to coal?

ANNA: Diamond.

TONY: You lose a life. What is the opposite of *hit*?

KATE: Miss. And I think the taboo letter must be M.

TONY: Correct. How did you guess?

KATE: *Malta, diamond,* and *miss* have only the letter M in common.

Compare Avoid 'em.

Alpha

- ∰ Any number of players

- ? Challenge game; word-finding game
- ✍ Played by writing or speaking
- ✎ Played with pencils and paper, or with no equipment

Object To think of words that start and end with the same letter.

Procedure Within a set time-limit, players have to write down as many words as they can that start and end with the same letter (e.g. *agenda, maxim, tight*). The player with the longest list is the winner. To make the game harder, players can be required to write down one word for each letter of the alphabet: *amoeba, bib, clinic,* etc. Extra points can be awarded for the players who think of the longest words for each letter.

Also called Bookends; Fore and Aft; Heads are Tails.

Alphabent

👥 Any number of players	✍ Played by writing or speaking
❓ Alphabetical game; challenge game	✎ Played with pencils and paper, or with no equipment

Object To write a sentence of 26 words, each word starting with successive letters of the alphabet.

Procedure You can try to do this on your own or as a timed competition with other people. The sentence can be written down or spoken aloud. If played by more than one person, successive letters can be called out in turn by each player (in which case the game may be called Oral Alphabent).

Example A boy can do extremely funny grimaces—holding in jaw, keeping lids minutely nearly open, peering quite rudely (sarcastically?) through, until very weary, X-pending youthful zeal. (Some players may think that the X-word is cheating.)

Alphabet

👥 Any number of players	✍ Played by writing or speaking
❓ Alphabetical game; challenge game	✎ Played with pencils and paper, or with no equipment

Object To think of a series of words which are prefixed by letters of the alphabet in order.

Procedure Players try to think of words like 'A-level' which start with a letter of the alphabet followed by a hyphen. The challenge is to build up a complete sequence of 26 such words, working through the alphabet. For example, for a start: *A-level, B-movie, C-stream, D-Day, E-type* . . .

There are two similar challenges involving the alphabet. One is to think of words that sound like all the letters of the alphabet: *Aye, Bee, Sea, Dee,* and so on. The other is to compile a sequence of words that would be thoroughly misleading if used when spelling out a word—for example, on the telephone. Sensible people spelling out a word like *page* would say: 'P as in *potato*, A as in *age*, G as in *get*, and E as in *eel*.' But a thoughtless person might say: 'P as in *psychic*, A as in *aesthetic*, G as in *gnome*, and E as in *ewe*,' completely confusing the listener. An intriguing pastime is to try thinking of a whole alphabet of similarly misleading spellings.

Alphabet Dinner

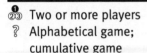

⨀ Two or more players
❓ Alphabetical game; cumulative game

✍ Played by speaking
✎ No equipment needed

Object To build up an alphabetical list of things to eat.

Procedure The first player starts by saying something like: 'Today I had for dinner some . . .' and names a food beginning with A. The second player says: 'Today I had for dinner some . . .', repeating what the first player said, and adding a food beginning with B. And so on, round the players. Anyone who cannot think of a suitable food or cannot remember the whole list is out of the game. The list may build up to something like this: 'Today I had for dinner some *apples, bananas, carrots, dumplings, escalopes,* and *frogs' legs*.'

11

Also called Alphabet for Dinner; Alphabetical Dining;
A-to-Z Banquet.

Compare Grandmother's Trunk; Hypochondriac; I Packed My Bag;
I Went to Market; Traveller's Alphabet.

Alphabet for Dinner *see* Alphabet Dinner.

Alphabetical Adjectives

② Two or more players	✍ Played by speaking
❓ Alphabetical game	✎ No equipment needed

Object To think of adjectives in an alphabetical series.

Procedure The first player says aloud a combination of an adjective
beginning with A and a noun, such as *attractive person*. The next player
then has to say a combination consisting of an adjective beginning
with B and a noun that starts with the same letter as the first noun,
such as *beautiful pig*. And so on, round the players. Anyone who
cannot think of a suitable phrase has to drop out of the game. The
game can also be played with successive adjectives starting with the
same letter and the nouns working through the alphabet, or both
adjective and noun starting with the same letter (*amiable artist*,
bewitched barmaid, etc.).

Compare One Awkward Albatross.

Alphabetical Dining *see* Alphabet Dinner.

Alphabet Race *see* Alphacross.

Alphacross

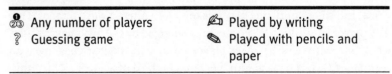 Two players	✍ Played by writing
? Grid game	✎ Played with pencils and paper

Object To use up all the letters of the alphabet in building a pattern like a crossword.

Procedure Each player takes a piece of paper and writes the 26 letters of the alphabet on it. A third piece of paper has a grid of blank squares drawn on it—say ten by ten squares (or you can use the empty grid from a newspaper crossword). The two players take it in turns to write words in the grid, interlocking with one another—as in a crossword. As they do so, they cross out from their own 'alphabet' the letters they have used. The first person to cross out all 26 letters is the winner. If neither player can use up all their letters, the winner is the one with the fewest letters left.

Example Kate writes *extinction* on the grid, and crosses out C, E, I, N, O, T, and X from her alphabet. She chose this word to get rid of X which, like Q and Z, is one of the hardest letters to use. Chas has to write a word that will interlock with *extinction* in some way, so he attaches *trickery* to one of the T's in *extinction*. He crosses out C, E, I, K, R, and Y from his alphabet (but not T, because that letter was already in the grid). And so on.

Also called Alphabet Race.

Compare Lynx.

Alphametics *see* Cryptarithms.

Alternade

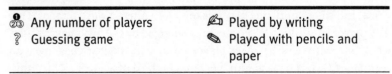 Any number of players	✍ Played by writing
? Guessing game	✎ Played with pencils and paper

Object To guess a word from other words made of its letters.

Procedure An alternade is a verse in which are hidden three words for players to find. Two of the words are made by taking alternate letters of the third word, which is called the keyword. So, if the keyword is *schooled*, the two other words would be *shoe* and *cold*. In the verse, the keyword is replaced by the word ALL or WHOLE; the two other words are replaced by ONE and TWO (or FIRST and SECOND). In more complicated alternades, the keyword can be divided in other ways: for example, into alternate *pairs* of letters.

Example

> The ghost ONE frighten all of us
> By rattling of his ALL,
> Which used to cause him quite a fuss—
> With both TWO legs tied to the wall.

Solution
ONE = *can*; TWO = *his*; ALL = *chains*.

Anaghost

👥 Two or more players	✍️ Played by writing or speaking
❓ Anagrams game	✏️ Played with pencils and paper, or with no equipment

Object To build up letters one at a time without making a word or an anagram.

Procedure This is a variation of Ghosts and Superghosts in which players in turn give letters which must *not* build up into a complete word. In this game, the letters must also not build up into an *anagram* of any word. The first player thinks of a word of three or more letters, and calls out its first letter. The second player adds another letter which continues but does not complete either a word or the anagram of a word, and so on, until one player is forced to finish a word or anagram. Any player who adds a letter can be challenged by the next player to say what the word will be. Any player who loses such a challenge, or completes a word or anagram, becomes 'a third of a ghost'. When this player loses again, he becomes 'two-thirds of a ghost' and the third time he is 'a whole ghost' and has to drop out of the game.

Example

TONY: R.
ANNA (*thinking of* for): F.
KATE (*thinking of* horrify): R.
CHAS (*thinking of* firer): E.
TONY (*thinking of* former): O.
ANNA (*challenges*): R, F, R, E, O is an anagram of *frore*, a word that
     means 'frosty'. (They check it is correct in the dictionary, and
     Tony becomes 'one-third of a ghost'.)

Anagrams (1)

Two or more players	🖎 Played by writing
❓ Anagrams game	✎ Played with pencils and paper

Object To solve anagrams.

Procedure Someone shuffles the letters of words, and gives the lists to
the players to rearrange into words. Usually the words are chosen from
a particular category—such as animals, flowers, or cities—and players
are told the category. The winner is the first player to rearrange all the
words correctly, or the one who solves the largest number of words in a
given time. Alternatively, players shout out the answer as they solve an
anagram: if correct, they are awarded a point and the players
concentrate on solving the remainder.

Example Chas prepared this list of anagrammed words for the others
to solve (the category is 'animals'): GRITE, WHEAL, YUPPP, BRAZE,
OBBING, ALEDROP, AAADKRRV, AEEHLNPT, MY ODD REAR,
and STEWIBLEED.

Also called Jumbled Words.

Solution *tiger, whale, puppy, zebra, gibbon, leopard, aardvark, elephant,
dromedary,* and *wildebeest.*

Anagrams (2)

Object To solve anagrams set by other players.

Procedure The first player thinks of a word that has an anagram, and says it to the second player, who tries to think of the anagram. If the second player succeeds in solving the anagram, that player presents another anagram to the next player. If the second player fails to guess the anagram, that player drops out for the round (or for the rest of the game), and the next player presents an anagram, and so on round the group. If desired, the first round can consist of three-letter words, the second round of four-letter words, and so on—to make the game increasingly difficult.

Example

KATE: Let's start with three-letter words. (*to Chas*): Bat.
CHAS: Tab. (*to Anna*): Act.
ANNA: Cat. (*to Tony*): Add.
TONY: Dad. (*to Kate*): Yam.
KATE: May. Now it's four-letter words. (*to Chas*): Came.
CHAS: Mace. (*to Anna*): Read.
ANNA: Dear. (*to Tony*): Dole.
TONY: Er... I don't know.
ANNA: Lode! You're out!

Analogies *see* What is My Thought Like?

Anapoems *see* Vocabularyclept Poetry.

Anaquote

Object To find a quotation which has been divided into sections.

Procedure One player thinks of a short quotation and divides it into groups of three letters. These groups are arranged in alphabetical order and presented to the other player or players, who try to reconstruct the quotation. If desired, a clue to the length of the words and the author's name can be given at the end, in the form of numbers representing the letters.

Example Tony presents the other players with this series of letter groups: ARI, BOT, EUN, FTH, GEO, GUA, HEA, LAN, OTI, RD, SAT, THE, TOM (1 4 2 2 6 3 8 2 3 7, 6 6 4). The letters RD are the last two letters of the quotation. The first ten numbers indicate the letters in each word of the quotation; the last three numbers indicate the letters in the author's name.

The other players juggle the letters around, and Kate eventually writes down the correct quotation and its author: 'A riot is at bottom the language of the unheard' (Martin Luther King).

Anguish Languish

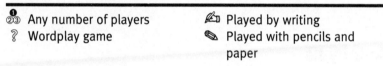

 Any number of players   Played by writing
 Wordplay game   Played with pencils and paper

Object To rewrite pieces using homophones.

Procedure The aim of the game is to reword pieces of writing, using different words that sound roughly the same. Thus, 'What a grey day' can be turned into 'What a grade A' and 'Underestimated' can be turned into 'Undressed, he mated'. Players can try to make up a whole story or poem like this.

Example

> *Song*
> My body-lice suffer devotion,
> My body-lice suffer disease,
> My body-lice suffer devotion.
> Oh, bring back my bonnet. Who, me?

Also called Holorimes; Junctures; Ormonyms.

Animal, Vegetable, and Mineral *see* Twenty Questions.

Animal, Vegetable, or Mineral *see* Clumps.

Antigrams

② Any number of players	✍ Played by writing
? Anagrams game	✎ Played with pencils and paper

Object To make anagrams which mean the opposite of the anagrammed words.

Procedure The challenge in an antigram is to rearrange the letters of a word so as to make a word or phrase that means the opposite. For example, *violence* can be turned into *nice love*, *funeral* into *real fun*, and *militarism* into *I limit arms*. Players can decide for themselves whether there are opposite meanings in the pairs *evangelists—evil's agents* and *prosecutor—court poser*.

Antonyms

② Two or more players	✍ Played by speaking
? Word-finding game	✎ No equipment needed

Object To think of words that mean the opposite of given words.

Procedure An antonym is a word that means the opposite of another word. *Good* is an antonym of *bad*, and *seldom* is an antonym of *often*. Antonyms can be used in two different games. Players can say a word and ask other players to think of a word that means its opposite. The first person to think of an acceptable antonym becomes the one to set a new word for the other players. A more complicated game is for people in turn to ask the other players to think of a word that means the opposite of *two* words. For example, which word is the opposite of both *observe* and *mend*? The answer is *break* (in such phrases as *break a promise* and *break a plate*).

Also called Opposites.

Compare Antonyms and Synonyms; Synonyms.

Antonyms and Sentonyms *see* Antonyms and Synonyms.

Antonyms and Synonyms

❶ ②③ Two or more players	✏️ Played by speaking
❓ Word-finding game	✎ No equipment needed

Object To think of words that mean the opposite of, or the same as, given words.

Procedure The first player says a word—preferably an adjective. The second player then has to say an antonym: a word that means the opposite. The next player says a synonym: a word that means the same as the preceding word. The next player says an antonym of this. And so the play proceeds: antonyms alternating with synonyms. Anyone is out if they cannot think of an appropriate word fairly quickly.

Example

ANNA: Sweet.
KATE: Sour.
CHAS: Grumpy.
TONY: Cheerful.
ANNA: Happy.
KATE: Sad.
CHAS: Funereal.
TONY: Er ... (*He has to drop out.*)

Also called Antonyms and Sentonyms.

Compare Antonyms; Synonyms.

Armadillo *see* A What?

Arrow of Letters

❶ ②③ Two players	✏️ Played by writing
❓ Word-building game	✎ Played with pencils and paper

Object To build up words, starting with a sequence of four letters.

Procedure One after the other, each player writes a letter inside a circle in a straight line, until a four-letter word is constructed. The circles are connected by arrows, and the four letters must be different from one another. The players then alternately add a previously unused letter inside another circle and indicate by arrows the word or words that have been constructed. Players score one point for each letter in the words they have created. The arrows may not cross one another, and words can only be made in the direction of the arrows. No more than four arrowed lines can be attached to any letter and no two letters can be joined by more than one arrow. Players cannot claim for a word that is included within another (e.g. if you make P–L–E–A–S–E, you can score five points for it, but nothing for *plea* or *peas*). The game ends when neither player can add any more word-making letters.

Example Kate draws T in a circle, Chas adds O, Kate adds Y, and Chas adds S, creating this:

Kate adds the letter A to make the word *soya*, scoring four points thus:

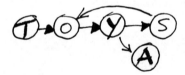

Chas wants to add Y to make *stay* but he can't because this letter has already been used. So he adds B to make *stab* and score four points:

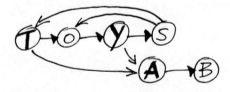

After several more moves, the diagram might look like this:

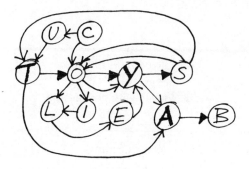

Also called Verbal Sprouts.

Background Invented by Michael Grendon and published in *Games and Puzzles* (1975).

Association Chain *see* Associations.

Associations

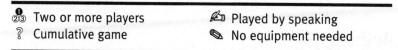

①②③ Two or more players	✍ Played by speaking
? Cumulative game	✎ No equipment needed

Procedure The first player says a word. The second player says the first word that comes into his or her mind as a result of the preceding word. And so on, round the players. The enjoyment lies in the sometimes unexpected sequence of words, and the hints they give about the players' personalities. To make it more difficult, at a certain point the players stop adding words and try instead, one by one, to retrace the chain backwards. This is sometimes called Association Chain.

Alternatively, the game can be played so that a word spoken must deliberately have *no* association with the preceding word. For this game, see Word for Word.

Also called Chain of Suggestions; Free Association; Pelmanism; Suggestion Chain; Word Associations; Word Order.

Compare Stepping Stones; Tennis, Elbow, Foot; Word for Word.

A-to-Z Banquet *see* Alphabet Dinner.

Autantonyms

Any number of players	Played by writing or speaking
Challenge game; word-finding game	Played with pencils and paper, or with no equipment

Object To find words which have two opposing meanings.

Procedure Autantonyms are words which have two contrary meanings, like *fast*—which can mean 'moving quickly' and also 'not moving' (as in 'stand fast' or 'fast colours'). Autantonyms can be used as a game in two ways. Players can give clues to the two different meanings of a mystery word, from which the other players try to guess the word. Or players can make as long a list of autantonyms as they can in a set time.

Author and Title *see* Awful Authors.

Authors

Three or more players	Played by speaking
Challenge game	No equipment needed

Object To tell a story in the style of a particular author.

Procedure One player is chosen to be the umpire, directing the game. The other players each choose an author (or kind of book) in whose style they will tell part of a story. The umpire gives them a title for a story, and the first player starts telling it, in the style chosen by that player. At a certain point in the story, the question-master tells the next player to continue the story in the style chosen by *that* player, and so on, round the players.

Example Anna is chosen as umpire and asks the other players what style they will use. Tony chooses the *Oxford English Dictionary*; Kate chooses an elementary reading-book; and Chas chooses Noel Coward. Anna asks them to tell the story of *Goldilocks and the Three Bears*.

TONY: *Goldilocks*, a species of young girl characterized by long, blonde hair and insatiable curiosity. *Three*, one more than two. *Bears*, carnivorous ursine mammals with thick hair, big feet, and . . .

ANNA: Over to Kate to continue the story.

KATE: The house is big. The house is red. The house is empty. Goldilocks enters the house. The house has a table. There are plates on the table. Goldilocks sees the table. Goldilocks sees the plates. Goldilocks sees the porridge. Goldilocks tastes the porridge . . .

ANNA: Enough! Over to Chas.

CHAS: 'Very flat, this porridge,' said Goldilocks. 'And extremely hot. In fact, damnably hot. It's as hot as the noonday sun into which only mad dogs and Englishmen venture. In fact, I seldom eat porridge for breakfast: kippers and kedgeree is about all a civilized person can face at this time in the morning. But it is extraordinary how potent cheap porridge can be . . .'

Automynorcagrams

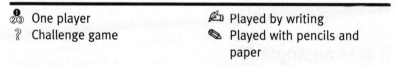

🔵②③ One player   ✍ Played by writing
❓ Challenge game   ✏ Played with pencils and paper

Object To write a sentence in which the first letters of the words spell out the beginning of the sentence. So the second word starts with the second letter of the sentence; the third word starts with the third letter of the sentence; and so on.

Example The house Emily Hayes occupied until she eventually encountered Maurice is luxurious. Yes, houses . . .

Avoid 'em

Two or more players
? Challenge game

✐ Played by speaking
✎ No equipment needed

Object To answer questions without using words containing a specified letter.

Procedure One player is chosen as umpire and states a letter of the alphabet which the other players must avoid. The umpire then asks each player in turn a question, which must be answered using words that do not contain the forbidden letter. Any player who uses the forbidden letter has to drop out of the game. Obviously the umpire can choose questions which may lead to words containing that letter.

Example Kate is chosen as umpire and tells the players that they must avoid the letter T.

KATE (*to Anna*): What do you drink at teatime?

ANNA: I usually have lemonade.

KATE (*to Chas*): How many are there in a dozen?

CHAS: One more than eleven.

KATE (*to Tony*): What is philosophy?

TONY: It's the study of . . . er . . . logic.

KATE: You have used three T's in your answer, so you are out!

Compare Aesop's Mission.

A Was an Apple Pie

Two or more players
? Alphabetical game;
cumulative game

✐ Played by speaking
✎ No equipment needed

Object To build up an alphabetical sequence.

Procedure The first player starts by saying: 'A was an apple pie' and then must add a verb beginning with A, for example: 'A ate it.' The next player has to think of a verb beginning with B, for example: 'B baked it.' And so on, round the players, working through the alphabet. Players who cannot think of a suitable word have to drop out, until there is only one player left: the winner.

Example

CHAS: A was an apple pie. A ate it.
KATE: B baked it.
ANNA: C cut it.
TONY: D dephlogisticated it.
CHAS: Pardon? E extruded it.
KATE: F forgot it ... (*And so on.*)

Awful Authors

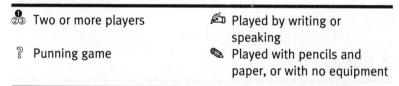

① ②③ Two or more players	✍	Played by writing or speaking
❓ Punning game	✎	Played with pencils and paper, or with no equipment

Object To make up punning names for the authors of imaginary books, such as '*The Broken Window* by Eva Brick'.

Procedure One player thinks of a book title, and gives it to the other player(s) to supply a suitably punning author. If the player giving the title has a particular author in mind, he or she can give the author's forename as well as the book title.

Example

ANNA: Who wrote this book: *Explosives*?

TONY: Dinah Mite?

KATE: T. N. Tee? What is the surname for the writer of this book: *The Unsatisfactory Bath* by Luke . . .?

CHAS: Luke Warm. What is the surname for the writer of this book: *The Lion Tamer* by Claude . . .?

TONY: Claude Body?

KATE: Claude Bottom?

Also called Author and Title; Game of the Name; Titles and Authors; Who Wrote What.

A What?

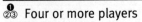

Four or more players	Played by speaking
Active game; cumulative game	Played with objects to be passed round

Object To pass an object round a circle with questions and answers.

Procedure The players sit in a circle and the first player passes an object (a ball, a cushion, etc.) to the person on his or her left, saying: 'This is a cat.' That player has to say: 'A what?' and the first player repeats: 'A cat.' The second player then passes the object to the person on his left, repeating 'This is a cat'. The third player asks 'A what?' but by this time the second player has forgotten what it was, so he turns to the first player and asks again 'A what?' The first player says: 'A cat.' The second player passes on this information to the third player. The object continues on its way round the circle, with the same pattern of questions and answers. When the object is about halfway round the circle, the first player passes another object to the person on his or her *right*, saying something like: 'This is an armadillo' and the same question and answer are again repeated. When the two objects meet and pass one another, there is bound to be some confusion - but that is the pleasure of the game.

Also called Armadillo; The Cat and the Dog.

Compare My Name is Mary.

B

Backenforth

② ③ Two or more players	✍ Played by speaking
? Spelling game	✎ No equipment needed

Object To guess words when they are spelt backwards.

Procedure One player spells out a word backwards. The first player to identify the word correctly scores one point. If players guess the word incorrectly, they lose a point. Alternatively, the person who guesses correctly becomes the next player to spell out a word backwards. The game can also be played with one player *pronouncing* a word backwards (e.g. 'taerg' for *great* backwards).

Compare Backward Spelling Bee.

Backward Spelling *see* Backward Spelling Bee.

Backward Spelling Bee

② ③ Two or more players (or two teams)	✍ Played by speaking
? Spelling game	✎ No equipment needed

Procedure One player (or a member of one team) asks another player (or a member of the other team) how to spell a word—but to say the letters backwards. If desired, the words can gradually increase in difficulty: starting with three-letter words, then moving to longer and longer words.

Example

ANNA: How do you spell *sausage?*
TONY: E-G-A-U-S-A-S.
ANNA: Wrong! It's E-G-A-S-U-A-S.

Also called Backward Spelling; Belling Spee; Spellbound; Spelling Crab.

Compare Backenforth.

Backwords *see* Reversals.

Bacronyms *see* Reversals.

Bananas *see* Sausages.

Banishments *see* Dismissals.

Battle Words *see* Word Battleships.

Beheadings *see* Beheadments.

Beheadments

❶ **②③** Any number of players	✍ Played by writing or speaking
？ Anagrams game; letters game	✎ Played with pencils and paper, or with no equipment

Object To find words which differ by one letter.

Procedure This game can be played in two ways:

1. Clues are given to a pair of words, in which the second word is formed by removing one letter (often the first letter) from the first word (as in *place* and *lace*). So one player might tell another player: 'Behead a woodwind instrument and get a stringed instrument.' The answer is *flute–lute*. When the last letter of the word is to be removed, the game is called Curtailments.

2. Players are given a set time to write down as many words as they can which can be 'beheaded' to form new words. Sometimes the challenge is to write down one such word beginning with each

letter of the alphabet. The winner is the player with the longest list of words, or the player whose words contain the largest total of letters.

Also called Beheadings; Decapitations; Remove a Letter.

Compare Shrink Words; Transdeletions.

Belling Spee *see* Backward Spelling Bee.

Bing-Bangs *see* Bullets.

Birds, Beasts, Fishes, or Flowers

① ②③ Four or more players		✍ Played by speaking	
? Active game; guessing game		✎ No equipment needed	

Object To guess a word from its initial letter.

Procedure The players form into two teams and line up facing one another from either side of a large room, open space, etc. The players on one side agree on a type of bird, beast, fish, or flower, and then advance in line abreast to face the other team. They state the initial letter of the chosen word. If the other side guesses it correctly, the first team turns and runs back to its original position, being chased by the second team. Anyone who is caught has to join the second team, which then chooses another mystery word. If the second team cannot guess the word, the first team chooses another word.

Black Squares

① ②③ Two players		✍ Played by writing	
? Grid game		✎ Played with pencils and paper	

Object To fill a grid with letters and to find 'black squares' which cannot contain any letter that makes a word with adjoining letters.

Procedure The players draw a grid, usually 12 by 12 squares. They then alternately enter letters into the spaces in the grid—any number of letters so long as they make a word (horizontally or vertically) or can be made into words by adding more letters. Players also try to find 'black squares' which cannot be filled by any letter that makes a word with adjoining letters. One point is scored for each 'black square' that a player finds, and two points for any opponent's 'black square' that can be successfully challenged by being filled with an acceptable letter. A player claims a black square by drawing a line through it. If the square is successfully challenged, the line is erased. If the square cannot be challenged, its claimant blacks it out completely. The game ends when neither player can add any more letters—or when neither player *wants* to add more letters, because it would give the other player a winning number of black squares. In some versions of the game, the first player starts by writing a letter in the *centre* square.

Example Tony starts the game by writing the letter P in a square. Anna writes the letter N in the square two along from the P. Tony does not challenge it, because he can see that the word *pin* or *pan* can be made (or even *Spanish*). He writes an S above the P (as many words begin with SP). Anna writes an E next to the N of P–N, knowing that she can make words like *pane* or *pine* or *supine* if she wants to. And so the players continue, carefully avoiding completing whole words, in case their opponent claims a 'black square' before or after that word.

Compare Pi.

Blends *see* Portmanteau Words.

Blockbusters

Two or more players	✍ Played by speaking
? Guessing game	✎ Played with a prepared set of questions and a grid of squares each containing one or more letters

Object To guess words from their initials.

Procedure Players have to 'bust' a path through a 'block' or grid of 20 interlocking squares or hexagons, each one inscribed with a letter or letters. A player chooses one of the squares and has to solve a question whose answer has the initial letter(s) shown in that square. For example, a player may choose a square marked S and the question-master asks: 'Which S is a brilliant red colour?' The answer is *scarlet*. If the square is marked MLK, the question might be: 'Which MLK was a famous Civil Rights leader in the USA?' and the answer would be *Martin Luther King*. The winner is the first person or team to break through the 'block' of squares by guessing the answers for five squares in a row. If one player or team cannot identify the letter(s) in one square, the opposing player or team has the chance to choose a square to guess. The layout of the 'blocks' enables players to choose an alternative route if they cannot answer a particular question.

Example A typical 'block' looks like this:

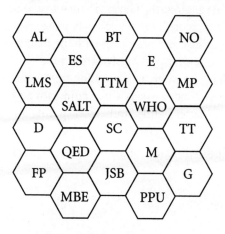

A player might choose the hexagons marked AL, ES, TTM, E, and MP. If this player correctly guesses the answers (which might be: Abraham Lincoln, El Salvador, *The Third Man*, Elsinore, and Member of Parliament), a clear path is made from left to right through the 'block' and this player wins the round.

Background A long-running game show (on Independent Television), played by teenage students and hosted by Bob Holness. A trade mark of Central Independent Television.

Boggle

 Any number of players

? Grid game; word-building game

 Played by writing

✎ Played with a tray into which 16 cubes are shaken, plus pencils and paper

Object To make up words, using letters appearing on the upper faces of cubes.

Procedure The 16 cubes are placed in a dome-shaped container, and a tray is placed over the container. The tray contains 16 spaces into which the cubes drop as the container is turned over and shaken. Within a set time, players try to make words (of three or more letters) from letters which join one another in the resulting grid: horizontally, vertically, or diagonally. No cube may be used more than once in making a word. When the set time expires, the players in turn read out the words they have made. They score points for each word that nobody else has found. The winner is either the player with the highest total of points or the first player to reach a previously agreed total. The official method of scoring is that one point is scored for three- and four-letter words, two points for five-letter words, three points for six-letter words, five points for seven-letter words, and eleven points for longer words.

Background Boggle is a registered trade mark of the Tonka Corporation.

Boloney *see* Sausages.

Bookends *see* Alpha.

Botticelli

 Two or more players

? Guessing game

 Played by speaking

✎ No equipment needed

Object To guess an identity adopted by one player.

Procedure One player thinks of a person to represent, and tells the other players the first letter of the person's name. The players try to guess the character by asking questions. The questions must expect a specific answer. So, if the character's name starts with B, a questioner might ask: 'Are you a jazz pianist?' and the answerer could reply: 'No, I am not *Count Basie*.' If the answerer cannot think of a jazz pianist, he or she can then be asked a general question, like 'Are you living or dead?' Whoever guesses the character correctly becomes the next answerer. If nobody guesses within a certain time-limit (or a specified number of questions), the answerer can choose a new identity.

Example Kate thinks of a person and tells the others that the initial is S.

ANNA (*thinking of* Shakespeare): Are you a famous writer?

KATE: No, I am not *Bernard Shaw.*

CHAS (*thinking of* Stockhausen): Are you a composer?

KATE: No, I am not *Sibelius.*

TONY (*thinking it may be* Barbra Streisand): Are you someone who had a hit record entitled *Memory* in 1982?

KATE: I don't know who you mean.

TONY: All right, are you a man or a woman?

KATE: A woman. (*And so on, until someone guesses that the personality is* Marie Stopes.)

Also called Garibaldi; Who Am I?

Compare Charades; Crambo.

Bouts-rimés

<table>
<tr><td>⊕
②③ Any number of players</td><td>✍ Played by writing or speaking</td></tr>
<tr><td>❓ Challenge game; poetic game</td><td>✎ Played with pencils and paper, or with no equipment</td></tr>
</table>

Object To write a poem using previously arranged words at the end of each line.

Procedure One player gives the others a set of words, usually an even number of rhyming words. Alternatively, the players can each contribute a word. (If playing alone, a player can choose words at random from a dictionary or other book.) Within an agreed time, each player has to write or recite a poem using those words at the ends of consecutive lines. If the poems are written down, the players read their own efforts aloud.

Alternatively, the first player can say a line and the second player must follow with a second line that rhymes, and so on round the players.

Example The chosen words are: *day, may, hill,* and *still.* Kate writes a humorous poem:

> It was a sunny, fine spring day,
> Quite early in the month of May,
> When Jack and Jill climbed up the hill—
> I wonder if they're up there still!

Tony's poem is rather more serious:

> 'What lies beyond the horizon?' I asked myself one day,
> Made up my mind to travel off, and find out - come what may.
> I crossed the horizon by climbing the hill,
> And looked—and saw—horizon still.

Also called Quatrains; Rhymed Endings; Rhyming Ends.

Build-Up

⚙ Two or more players   ✍ Played by writing
❓ Word-building game   ✎ Played with pencils and paper

Object To build up words from groups of letters.

Procedure Each player is given a previously prepared list of twelve groups of letters: four groups of four letters, four groups of three letters, and four groups of two. Alternatively, the players can build up the list by each suggesting a group of letters in turn. Each player then tries to write down a list of the longest words they can think of that each contain one of these letter-groups somewhere in the middle—one word for each group of letters. So that, if one of the groups of letters is AND, one player might write down *scandalous* and another might write *pandemonium*. At the end of a given time, the players add up the total number of letters in the words they have written, and the winner is the player with the highest total. Double points are scored for any words that contain the same group of letters twice, and triple points are scored for words that contain a letter-group three times. The game can be simplified by using a smaller number of letter-groups. Alternatively, just one word or prefix can be chosen, for players to see how many words they can make that start with that word or prefix.

Example The four players in turn suggest three groups of letters, resulting in the following list:

ATTE	ANT	ER
GEST	IGE	IN
NGEL	REM	OL
UNCH	YON	TU

From this list, Anna makes the following words (with scores indicated):

FLATTER	7
DOMINANTLY	10
SERIOUS	7
DIGESTION	9
TIGER	5
SINGER	6
CHANGELING	10
GREMLIN	7
POLTROON	8
TRUNCHEON	9
MAYONNAISE	10
STUDENT	7
	95

But she is beaten by Kate, who makes the following words:

FLATTEN	7
PANTALOONS	10
BERBERIS	16
SUGGESTIVE	10
PIGEON	6
TINTINNABULATION	32
ANGELIC	7
TREMBLE	7
COLLISION	9
STAUNCHNESS	11
BEYOND	6
FLATULENT	9
	130

Note that Kate gets a double score for the words *berberis* and *tintinnabulation*, as they both include their letter-group twice.

Also called Combinations.

Compare Superghosts.

Bullets

⚉ Any number of players

? Challenge game

✍ Played by writing or speaking

✎ Played with pencils and paper, or with no equipment

Object To think of appropriate phrases following particular patterns.

Procedure Players choose, or are given, a short phrase (usually of two or three words) and try to think of another short phrase which describes or says something about the original phrase. In various versions of the game, there are rules about the words that may be used; the words must: (1) start with the same initial letters as the words in the original phrase (for example, *A Sponge*—Absorbing Subject), or (2) start with the first letter and the last letter of the original phrase (for example, *William Shakespeare*—Wrote English), or (3) use a rhyming word or words (for example, *Good Boy*—Parents' Joy).

Also called Bing-Bangs; Camouflage; Cutlets; Heads and Tails; Mascots; Nuggets; Phrases; Simplets; Trumps; Winners.

Background Started as 'Dittoes' in *Tit-Bits* magazine in about 1909 but copied by many other magazines, particularly *John Bull*, which called it 'Bullets'.

Buried Words *see* Hidden Words.

Buzz

⚉ Three or more players

? Challenge game; cumulative game

✍ Played by speaking

✎ No equipment needed

Object To substitute words for particular numbers in counting.

Procedure The players sit in a circle. The first player calls out 'One,' the next player 'Two,' and so on—counting round the circle. When any player reaches five, or a multiple of five (such as 10 or 15), that player has to say 'Buzz' instead of the number. If a number includes a five but it is not a multiple of five, the player has to replace that part of it with 'Buzz'—for example, 54 would be 'Buzz-four'. Any player is out of the game who says the wrong thing or hesitates too much.

The game can also be played with some other number than five being replaced by 'Buzz'. And it can be made more difficult by simultaneously substituting another word for another number, so that (for example) the number seven would have to be said as 'Fizz' or 'Bizz'. The game might then be called something like Buzz-Bizz, Buzz-Fizz, or Fizz-Buzz, and counting would start like this: one, two, three, four, buzz, six, fizz, eight, nine, buzz, eleven, twelve, thirteen, fizz, buzz.

Also called Fizz.

Buzz-Bizz *see* Buzz.

Buzz-Fizz *see* Buzz.

C

Call My Bluff *see* Dictionary Game.

Camouflage *see* Bullets.

The Cat and the Dog *see* A What?

Catchphrase *see* Rebus.

Catchword

② Two or more players	✍ Played by speaking
❓ Word-building game	✎ No equipment needed

Object To think of words containing three chosen letters.

Procedure Players are given (by a question-master or by the other players) random sequences of three letters and, within a given time-limit (say, 30 seconds), have to think of as many words as they can starting with the first of the three letters and including the other two letters in order. In some forms of the game, the first letter may not *need* to be at the start of words, or the last of the three letters *has* to be at the end of words. Each player is given a separate set of three letters and they receive points for every word made, plus bonus points for the longest word made in one round of the game.

Alternatively, the players are given a grid of nine letters (with a vowel in the centre) and they have to choose one set of three letters (reading horizontally, vertically or diagonally, including the vowel) from which to make words, for instance:

H	T	Y
S	E	G
B	M	D

Example Kate is given the letters P–F–T and in 30 seconds she thinks of *perfect, parfait, prefect, profit, profanity,* and *profiteer.* She scores points for all the words except *profiteer,* which comes from the same root as *profit* (words from the same root are not allowed). Chas is given the letters T–N–C and he comes up with *tunic, tonic, technicality, tench, tincture, turbulence, temperance,* and *tercentenary.* He scores eight points for these words, plus a five point bonus for the longest word in this round (*technicality* or *tercentenary*).

Also called Licence Plate Game; Ultraghost.

Background Invented by Alasdair Adamson and used as the basis for a BBC Television quiz show. The TV version also includes games with anagrams and a 'four-second game' in which contestants are given ten different three-letter sequences at four-second intervals: they have to think of suitable words within those four seconds.

Categories

Two or more players	Played by writing
Word-finding game	Play with pencils and paper

Object To find words that begin with a particular letter (or letters) of the alphabet and that fit particular categories.

Procedure This game is the same as *Guggenheim* except that sometimes only one letter is chosen. Players have to find words beginning with this letter, to fit prearranged categories.

Centos

Any number of players	Played by writing
Challenge game; poetic game	Played with pencils and paper

Object To create a poem out of lines taken from different poems.

Procedure Players have to make up poems consisting of lines taken from existing poetry. You can try this on your own, or do it with other people and entertain each other by reading out your attempts. Centos work best when the chosen lines are well known to most people.

Example The Duke in Mark Twain's *Huckleberry Finn* makes a kind of cento with lines from Shakespeare, mangling Hamlet's famous soliloquy:

> To be, or not to be; that is the bare bodkin
> That makes calamity of so long life;
> For who would fardels bear, till Birnam wood do come to
> > Dunsinane,
> But that the fear of something after death
> Murders the innocent sleep,
> Great nature's second course,
> And makes us rather sling the arrows of outrageous fortune
> Than fly to others that we know not of.

Also called Mosaics; Mosaic Verse; Patchwork Poetry; Patchwork Verses.

Centurion

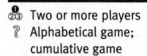

Two or more players	Played by writing
Alphabetical game; cumulative game	Played with pencils and paper

Object To make three-letter words that do not exceed a particular score.

Procedure The letters of the alphabet are given scores, so that $A = 1$, $B = 2$, $C = 3$, and so on to $Z = 26$. The first player writes down a three-letter word and, beside it, the total score made by its letters. The second player writes another three-letter word underneath it, starting with the last letter of the preceding word, and writing its score against it. No word may be repeated. The scores accumulate as the game progresses, and the loser is the first player who makes the score total 100 or more.

Double Centurion is played in the same way by three or more players, and the losing score is 200. Both games were invented by David Parlett.

Example Chas writes down FAD = 11 (6+1+4). Kate writes under it DAB = 18 (4+1+2, adding the 11 from the previous word). Chas writes BAD = 25 (2+1+4, added to the previous score of 18). Kate decides to push towards the end of the game and writes DUO = 65 (4+21+15+25). Chas thinks of writing OWE but realizes that this would take the score over 100, so he writes ODE = 89 (15+4+5+65). Kate cleverly writes EBB = 98 (5+2+2+89), which makes it impossible for Chas not to exceed 100 with his next move, so he loses.

Compare Inflation; Numwords.

Chain of Suggestions *see* Associations.

Chain Spelling *see* Ghosts.

Chain Story *see* Rigmarole.

Charades

⓪②③ Two or more players (usually two teams)	✍ Played by speaking or miming
？ Active game; guessing game	✎ No equipment needed (but players can 'dress up' if they wish)

Object To guess a word that occurs in, or is represented by, scenes acted by players.

Procedure The people present divide into two teams. One team goes out of the room and decides on a word to be acted out in syllables: the word must divide into syllables which are words in themselves—or sound like words. The team then acts a scene in which the first syllable is among the words spoken. The following syllables are 'acted' similarly, and then the whole word. If preferred, the scenes can be mimed, not spoken. If the other team guesses the word, it chooses the next word. If the other team fails to guess the word, the first team chooses another word to act out.

Example Chas and Kate leave the room, choose the word *goodbye*, and plan how they will act out its syllables. They come back into the room and firstly act out a scene in a schoolroom, in the course of which the teacher (Kate) tells the pupil (Chas) how *good* he is. They then act a scene in a shop, where a customer has a list of things to *buy* (the acted syllable can *sound* like the syllable to be guessed). Finally they include the whole word in a scene at a railway station, where two people are saying *goodbye* to one another. Anna and Tony correctly guess the word, and leave the room to choose one of their own.

Also called Spoken Charades.

Compare Crambo; Dumb Crambo; The Game.

Chesterfield

 Two players
? Grid game

✐ Played by writing
✎ Pencils and paper

Object To write down six chosen words before one's opponent.

Procedure The two players choose two related categories which are acceptable to them both: for example, trees and flowers. Each player writes down a secret list of six words in one of those categories. A three-letter word is chosen at random (e.g. from a book) and written vertically on a piece of paper (preferably paper divided into squares, like graph paper). The players then take turns to add words from their secret lists—at right angles, either vertically or horizontally—to the first or last letters of words which are already on the paper. If they cannot find a place for a word from their lists, they insert other words from their chosen categories. No word may be used more than once. The first player to write down all six of his or her chosen words is the winner. If a player cannot think of a word to write, the other player wins the game.

Example Kate and Anna respectively choose the categories 'trees' and 'flowers'. Kate writes down a secret list of six trees: *acacia, laurel, maple, poplar, rowan,* and *spruce*. Anna writes down these six flowers: *alyssum, crocus, daffodil, dahlia, primrose,* and *tulip*. From a book they choose at random the word *and*, which they write vertically on their sheet of paper. They each add words in their categories. When they cannot find places for words from their secret lists, they insert other trees or flowers. At the end of the game, the paper looks like this:

```
                    P
                    E
                    O
                    N
              L I L Y
              A
              U
              R
              E
D A F F O D I L
A
H                       P O P L A R
L        A C A C I A    R         O
I          N            I         W
A L M O N D    L        Y         A
               Y        M         N A R C I S S U S
               S        R             P
               S        O             R
               U        S             U
             M A P L E                C
                                      E
```

Kate is the winner because she has written down all six of her trees, while Anna has only managed to write down four of her flowers.

Background Invented by Peter Newby.

Chinese Gossip *see* Chinese Whispers.

Chinese Whispers

<table>
<tr><td>Three or more players</td><td>Played by speaking</td></tr>
<tr><td>Cumulative game</td><td>No equipment needed</td></tr>
</table>

Object To see how a phrase changes as it passes round a number of speakers.

Procedure The players sit in a circle, and the first player thinks of a phrase or sentence (a message, the title of a film or book, etc.) and whispers it into the ear of the next player. This second player whispers it to the third player and so on, round the circle. When it returns to the first player, that player announces what the phrase has become and how it started: the two versions are often wildly different.

Example

ANNA (*whispers to Kate*): I can't help eating jellies.
KATE (*whispers to Chas*): I can't tell Pete in wellies.
CHAS (*whispers to Tony*): I can tell Pete is smelly.
TONY (*whispers to Anna*): Ike and Tel's feet are smelly.

Also called Chinese Gossip; Russian Gossip; Russian Scandal.

Chronograms

 Any number of players
? Challenge game; guessing
game

✑ Played by writing
✎ Played with pencils and
paper

Object To use Roman numerals as letters in words or sentences.

Procedure Chronograms consist of one or more words in which letters also represent Roman numerals. The letters that can be used in this way are: C (100), D (500), I (1), L (50), M (1,000), V (5), and X (10). Occasionally the letter I is used for J and V is used for U, and two Vs can represent W.

The aim is to make a word, phrase, or sentence that includes an appropriate total if you add up the Roman numerals. For example, a book published in 1652 by Francis Goldsmith had his name spelt thus on the title-page: franCIs goLDsMIth, so that the capital letters add up to 1,652. And the year when Mount Everest was conquered (MCMLIII, that is 1953) is commemorated in the first letters of: 'Man Conquers Mountain's Last Incredibly Intriguing Impediment.'

Chronograms can be used as a game in several ways. Within a given time-limit, players can think of as many English words as they can that consist entirely of Roman numerals (for example, DIM, MILD, and CIVIL). Players can be challenged to write an appropriate chronogram for the year of their birth or the year of an important historical event. Or one player can present another with a problem to solve that includes Roman numerals—for example, which word is represented by this series of letters and figures?—
E1010001,0001,000UN1100 ATXN.

Also called Roman Numeral Words.

Solution

EXCOMMUNICATION.

Clubs *see* Clumps.

Clue Words

@ Two or more players   🖾 Played by speaking
? Guessing game   ✎ No equipment needed

Object To guess a word from clues which are anagrams of part of that word.

Procedure The first player thinks of a word of seven or more letters and gives the other players a clue to it, consisting of a three-letter word made of some of its letters (all different from each other). The other players in turn each have one guess at the word. Anyone guessing it gets three points, and chooses the next word for guessing. If nobody guesses it correctly, the first player gives another clue: a four-letter word made of some of the target word's letters. If anyone guesses correctly, they get two points. If nobody guesses correctly, the first player gives a five-letter word. The player who now guesses the target word correctly gets one point and chooses the next mystery word. If nobody guesses the word, the first player wins one point and the second player chooses the next mystery word. This game, probably invented by Gyles Brandreth, can be played with just two players, in which case they alternate choosing and guessing the word. To make the game easier or harder, a word of more or fewer letters can be chosen.

Example

ANNA: The three-letter clue is *roe*.

TONY: Is it *deplore?*

KATE: *Torture?*

CHAS: *Toaster?*

ANNA: No, none of those. The four-letter clue is *mole*.

TONY: *Volumes?*

KATE: It can't be: It's got to have an R in it. How about *remould?*

CHAS: *Polymer?*

ANNA: No, you are all wrong. The five-letter clue is *lover*.

TONY: So the word must contain L, O, V, E, R, and M. But I can't guess what it is.

KATE: I know! It's *removal!* (*And she is right.*)

Clumps

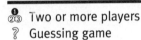 Four or more players
? Guessing game

✍ Played by speaking
✎ No equipment needed

Object To guess a word in teams by asking questions.

Procedure The players divide into two or more teams. One member from each team goes out of the room and together they decide on a word. When they return, the other members of their team cluster around them in a 'clump' and ask questions, which must only be answered *yes* or *no*. The first team to guess the mystery word is the winner. In some versions of the game, the two or more people who chose the mystery word are added to the team that guesses it correctly, and the winning team is the one that absorbs all or most of the players.

Also called Animal, Vegetable, or Mineral; Clubs.

Coffeepot

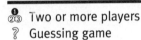 Two or more players
? Guessing game

✍ Played by speaking
✎ No equipment needed

Object To guess a word from sentences in which it is disguised as the word *coffeepot*.

Procedure One player thinks of a word that has two or more meanings (like *bat*), or two or more different words that sound the same (like *pair/pare/pear*). The other players then ask questions, which the first player must answer with a sentence including the chosen word (or one of the words)—but substituting *coffeepot* for that word. Any player who has just asked a question can try to guess the word. A correct guess scores a point, or allows the guesser to choose the next mystery word. If desired, the word *teakettle* or *teapot* may be used instead of *coffeepot*.

Example

CHAS: I have chosen my word.
ANNA: Do you like kippers?

CHAS: Only when I receive them in the coffeepot.
KATE: How intelligent are you?
CHAS: As intelligent as most coffeepots.
TONY: Why don't you get a job?
CHAS: I did at Christmas, working for the Royal Coffeepot.
TONY: Is the word *mail*?
CHAS: Yes—*mail* or *male*.

Also called Teakettle; Teapot.

Combinations (1)

⚫②③ Two or more players   ✍ Played by speaking
 ❓ Guessing game   ✎ No equipment needed

Object To guess words which make combinations.

Procedure One player thinks of a series of combinations (usually three) that have one word in common. This player then asks the other players to guess the common word by giving them the other words of the combinations.

Example Which single word can be added before all these words— *jam, light,* and *warden*—to make three combinations or phrases? The answer is *traffic,* making *traffic jam, traffic light,* and *traffic warden.* Which single word can be added *after* each of these words—*dog, light,* and *out*—to make three combinations or phrases? The answer is *house,* making *doghouse, lighthouse,* and *outhouse.*

Compare Connections.

Combinations (2) *see* Build-up.

Comparisons

⓪②③ Two or more players

? Guessing game

✍ Played by writing

✎ Played with a number of objects representing words, plus pencils and paper

Object To guess phrases from displayed objects.

Procedure One person prepares beforehand a number of objects which are laid out on a table. Each object is placed on a large numbered sheet of paper. The players are given pencils and paper and, within a given time, try to write down the phrases suggested by the objects. The objects are chosen to suggest comparisons, so a feather could represent the comparison *as light as a feather*, and a whistle could represent *as clean as a whistle*. Other suitable phrases include *dead as a doornail, as bold as brass*, and *as long as a piece of string*. The winner is the player who correctly guesses the largest number of phrases.

Concealed Words   *see* Hidden Words (1).

Concealments

⓪②③ Two or more players

? Word-finding game

✍ Played by writing

✎ Played with pencils and paper

Object To find words concealed in a chosen piece of writing.

Procedure Each player is given a copy of a piece of writing—from a book, magazine, or newspaper. The players have to find as many words as they can, hidden within the words of the piece or formed by consecutive letters.

Example One player chooses a passage from a story by James Thurber entitled *The Night the Bed Fell*: 'In case I didn't arouse him until he was about gone, he said, he would sniff the camphor, a powerful reviver.'

The players found the following words in this passage: *Inca* (in *In case*), *as* (inside *case*), *did, tar, rouse, us, use, munt* (a South African slang word), *hew, one, aid, if, he, camp, am, amp, rap, pow, power*.

Compare Proverb Delve.

Connections

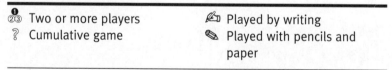

Two or more players	Played by writing or speaking
Guessing game	Played with pencils and paper, or with no equipment

Object To guess words that connect two phrases or combinations.

Procedure One player thinks of a pair of phrases or combinations that have one word in common, such as *teapot* and *pot luck*. This player then tells the other players the first part of the first combination and the second part of the second combination, asking them to guess the connecting word. Whoever guesses correctly is allowed to choose the next pair of words.

Example

KATE: Which word could connect *curtain* and *road*?

CHAS: *Rail*—which makes *curtain rail* and *railroad*. Which word connects *silver* and *ring*?

ANNA: *Wedding*. Which word might connect *village* and *light*?

KATE: *Green*.

Compare Combinations (1).

Consequences

Two or more players	Played by writing
Cumulative game	Played with pencils and paper

Object To create a story with separate contributions from each player.

Procedure Each player is given a sheet of paper, and starts by writing at the top an adjective describing a male person. The player folds over the paper to hide the adjective, and hands the paper to the next player, who writes a man's or boy's name. The papers are passed on from one player to another, each player writing in order the following:

(1) an adjective or adjectival phrase describing a man or boy
(2) a man or boy's name
(3) an adjective or adjectival phrase describing a woman or girl
(4) a woman or girl's name
(5) where they met
(6) what he said to her
(7) what she replied
(8) what the consequence was
(9) what the world said about it.

If desired, extra categories can be added between numbers 5 and 6: what he gave to her, what she gave to him, what he did, what she did, and when or how they met.

The players then read aloud the resulting stories, inserting linking words to make a continuous narrative.

As a variant of this game, a book or film review can be built up by the players writing:

(1) the title
(2) a sub-title
(3) the author's or stars' names
(4) the opening line
(5) a short review.

Example A game of Consequences might produce a nonsensical story like this: Underhanded Elvis Presley met voluptuous Mrs Gamp in the food hall at Harrod's. He said to her: 'Do you come here often?' and she replied: 'I only came in for a bunch of bananas.' The consequence was that they had a white wedding, and the world said: 'You don't get many of those to the pound.'

Compare Add a Word; Headlines; Pass It On; Rhyming Consequences; Rigmarole.

Constantinople *see* Words within Words.

Constructapo *see* Vocabularyclept Poetry.

Conundrums *see* Riddles.

Convergence

②③ Two players	✍ Played by writing
❓ Guessing game	✎ Played with pencils and paper

Object To guess a four-word sentence chosen by one's opponent.

Procedure Each player writes down a four-word sentence, concealing it from the other player. The sentences must not contain any proper names. One of the players tries to guess the other player's sentence by proposing a 'test' sentence of four words. The other player says whether these words alphabetically precede or follow the words in his or her hidden sentence. The second player then proposes a 'test' sentence, in an attempt to guess the first player's sentence. The players continue guessing in turn, until one of them correctly guesses the other player's sentence.

Example Chas is trying to guess Kate's sentence. He first of all says: 'My cat is lovely.' Kate replies: 'Before, after, before, before'—meaning that all the words in Chas's sentence, except the second word, come alphabetically *before* the corresponding words in her own sentence. Chas now proposes: 'Why buy white rabbits?' Kate replies: 'Before, after, after, before.' Chas now knows that the first word alphabetically follows *why*, the second precedes *buy*, the third is between *is* and *white*, and the fourth follows *rabbits*. He therefore guesses: 'You ate my teacake.' Kate says: 'Correct, same, before, before' (meaning that Chas has guessed the first word correctly, and he has the right initial letter for the second word, but he still needs to find words that follow *my* and *teacake*). Chas guesses: 'You are very useful' and eventually guesses correctly that Kate's sentence is 'You are no use.'

Compare Crash; Double Jeopardy; Jotto.

Countdown

⓪②③ Two or more players	✍ Played by writing or speaking
? Anagrams game	✎ Played with pencils and paper, or with no equipment

Object To make as long a word as possible from nine letters chosen at random.

Procedure Nine letters are chosen at random, and players have to make the longest word they can from those letters within a set time-limit (say, 30 seconds). In the television version of the game, the two contestants also have to solve a nine-letter anagram (called 'the conundrum', scoring ten points for a correct answer) and play a game with numbers. The person with the longest word scores one point for each letter in that word. If a nine-letter word is created, the player scores 18 points.

Example By sticking a pin in one page of a book, the following nine letters are chosen: A–T–E–N–O–R–I–M–L. From these, Tony makes the six-letter word *lament*; Kate and Chas each make seven-letter words (*ailment* and *latrine*); but Anna wins with the eight-letter word *terminal*. Nobody notices the possibility of using all nine letters to create *lion-tamer* (but the strictest rules would not allow hyphenated words).

Background Devised by Armand Jammot as the French game *Des chiffres et des lettres*, it was adapted for British television in 1982 and was the first show to be transmitted on Channel 4. Countdown is a registered trade mark of Yorkshire Television Ltd.

Crambo

 Two or more players
? Guessing game; rhyming game

✍ Played by speaking
✎ No equipment needed

Object To guess a word by means of rhyming words.

Procedure The players divide into two teams. One team goes outside the room, while the other team chooses a word—preferably a word that rhymes with several other words. The other team re-enters the room and is told a word that rhymes with the mystery word. They then try to guess it, and convey their guesses to the other team in clues, without mentioning what word they think it is. When they guess the word correctly, the other team leaves the room, to become the guessers. Alternatively, the game can be mimed instead of spoken (see Dumb Crambo). The game can also be played without teams, the players simply setting mystery words for one another to guess.

Example Kate and Tony leave the room. Anna and Chas choose the mystery word *bag*. Kate and Tony re-enter, and are told that the mystery word rhymes with *flag*.

TONY: Is it a rugged rock?

ANNA: No, it is not *crag*.

KATE: Is it to repeat unpalatable truths?

CHAS: No, it is not *nag*.

TONY: Is it a shortened name for a type of posh, fast car?

ANNA: No, it is not *Jag*.

KATE: Is it a cloth container?

ANNA and CHAS: Yes, it is *bag*.

Also called Acting Rhymes; Rhyming Tom.

Compare Botticelli; Charades; Dumb Crambo.

Crash

Two players

Guessing game

Played by writing

Played with pencils and paper

Object To guess a five-letter word chosen by one's opponent.

Procedure Both players write down a five-letter word for the other player to guess. The first player tries to guess the second player's word by suggesting a 'test' word of five letters. The second player tells the first which letters of the test word correspond to letters in the hidden word. The second player tries to guess the first player's word in the same way, and they continue alternately until one guesses the other's word correctly.

Example Anna writes down the word *potty*. Tony proposes *these* as his test word. Anna tells him that none of the letters in *these* corresponds with the letters in her chosen word. Tony then suggests *silly* (choosing a word with completely different letters) and is told that the fifth letter is correct. He goes on through *merry, perry, pasty,* and *patty* until he correctly guesses *potty*.

Also called Words.

Compare Convergence; Double Jeopardy; Jotto.

Croakers *see* Tom Swifties.

Cross-Questions

Eight or more players

Cumulative game

Played by writing and speaking

Played with pencils and paper

Object To create a series of ridiculous questions and answers.

Procedure The players divide into two teams, and each team chooses a leader. The leader of one team prepares a set of random questions; the leader of the other team—without consulting the other leader—prepares a series of answers. It is best if the leaders agree beforehand the form of the questions and answers, for example: questions starting 'Why...' and answers beginning 'Because...'.

The two teams form into two straight lines (either standing or sitting) and their leaders pass down these lines whispering a question or answer to each player. The first player in the 'questions' team then has to say his or her question out loud, to be answered by the first player in the 'answers' team. The fun of the game is in the often ludicrous pairings of questions and answers. For the next round, the questioners become the answerers, and vice versa.

Example The leaders agree to start their questions and answers with the word 'When...'. They then whisper their questions and answers to the players, resulting in this recitation:

QUESTION: When is a door not a door?
ANSWER: When I say so.
QUESTION: When is dinner?
ANSWER: When the red, red robin, comes bob, bob, bobbin' along.
QUESTION: When shall we three meet again?
ANSWER: When there's an 'R' in the month.

Also called Cross-Questions and Silly Answers.

Cross-Questions and Silly Answers *see* Cross-Questions.

Cross Wits

②③ Any number of players	✍ Played by writing or speaking
? Grid game	✎ Played with a previously prepared grid

Object To solve clues to six interlocking words.

Procedure Players have to solve clues to six words which interlock in a grid like a small crossword. The words have a linking theme and a clue is given to that theme as well as to each of the words. If the game is played by two teams, they try alternately to solve the clues. Players score one point for every letter in a word they solve, plus a bonus for guessing the 'theme' word.

Example

Clues

Theme: Drum.

- (1) It carries liquids in bulk.
- (2) A handy tree.
- (3) The first name of Popeye's beloved.
- (4) Clever—or dirty?
- (5) What's on your mind?
- (6) A ship's rope makes pretty pictures.

Solution

Theme: Oil.

Background Cross Wits is a television game-show produced by Tyne Tees TV for ITV. In the television version, there are two teams—each consisting of a celebrity and a member of the public. In the final round, one of the finalists has to guess ten clues in 60 seconds.

Crossword

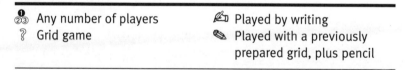

Any number of players	Played by writing
Grid game	Played with a previously prepared grid, plus pencil

Object To complete a puzzle by entering words on a grid in response to clues.

Procedure A crossword is a chequered diagram, usually square, in which the player has to enter words guessed from numbered clues. The words read 'across' (horizontally) or 'down' (vertically). The words are separated by black squares (or occasionally by thick bars between squares). Crossword grids are usually designed so that the pattern of black and white squares looks the same if the grid is turned upside down. The words interlock with one another, so that solving one word will supply one or more letters for other words. Easy or quick crosswords usually have synonyms as clues; harder or 'cryptic' crosswords have clues which lead indirectly to the answers, using such devices as anagrams, double meanings, homonyms, etc.

Example Tony starts doing a crossword puzzle in the newspaper, with Chas looking over his shoulder. It is a cryptic crossword, so the clues conceal the answers in various ways.

TONY: What's the first word? '1 across: Old man's weapon demanded by sentries (8 letters).' What does 'old man' mean? Is it Methuselah?

CHAS: You mustn't necessarily interpret this sort of crossword's clues literally. This looks like one of those words that is made up from parts of two or more words. Take it bit by bit. What else does 'old man' mean? It could be someone's husband.

TONY: Or someone's father: after all, I am your 'old man'. Father's weapon ...?

CHAS: Or it could be 'Pa's weapon'. Pa's gun? Pa's sword? That's it! Pa's sword—*password*. That is what sentries ask for.

TONY: Good. Now, '1 down: Peel's creation, initially (6, 9).' 'Initially' in a clue usually suggests there is an abbreviation somewhere. The answer starts with P, and the initials of 'Peel's creation' are *PC*, so it could be *policeman*, as the police were created by Sir Robert Peel. I know! *Police constable.*

CHAS: Right. Now what's this 12 across, seven letters: 'Disturbingly angered'? A word in a clue like 'disturbed' or 'upset' or 'shuffled' often signals an anagram. What is an anagram of *angered*?

TONY: *Enraged.* That gives us an R at the beginning of the second word in 6 down: 'The cylinder is jammed (5, 4).' Does that mean 'Your car won't work'?

CHAS: You are being too literal again. These cryptic clues often use double meanings, so it doesn't necessarily mean that a cylinder is literally jammed. What's another word for a cylinder?

TONY: Tube? Pipe? Roller?

CHAS: We know that the second word starts with R, so it could be *roll*. I know: *Swiss roll*, because it's a round thing with jam inside.

TONY: Very good. What about this clue: 'Part of the Clyde bristling with wreckage (6).' A word like 'part' or 'in' often signals a hidden word, so we could look for a word hidden in the clue.

CHAS: I can see it: *debris*, in 'Cly*de bris*tling'. That gives us an R to start 21 down: 'Terrible driver had piggy-back ride, we hear (4, 3).' Doesn't a phrase like 'we hear' or 'they say' or 'sounds like' suggest that the answer sounds like something else?

TONY: Yes. 'Had a ride' might be *rode*, which sounds like *road*, but what is this piggy-back? I know: the answer is *road hog* because that sounds like someone 'rode a hog' (a piggy)—and a road hog is a terrible driver.

CHAS: So are you.

Also called Crossword Puzzle.

Crossword Game

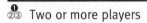

Two or more players	Played by writing
Grid game	Played with pencils and paper

Object To fill a grid with words made up from letters chosen by the players.

Procedure Each player draws a blank grid of squares—five by five, seven by seven, or larger. The players in turn call out a letter of the alphabet. As the letter is called, each player writes it into any square on their grid that they choose. As they do so, they try to build up words, reading across or down (or both). The players continue calling out letters until all the squares are full. Players score points for the number of letters in each word that they have made. Individual letters can score as many times as they are used in separate words.

Example The players used a grid of five by five. Tony's result looked like this:

H	O	F	T	K
E	D	I	M	S
A	C	L	U	B
R	E	J	D	G
T	H	O	S	E

For words 'across', Tony scores points for *dim, dims, club, those,* and *hose*; for words 'down' he scores for *hear, heart, ear, art, mud,* and *muds*: a total score of 42. Can you see any words he missed?

With the same letters, Anna's grid looked like this:

I	S	U	E	T
F	O	R	A	M
H	L	G	C	B
O	D	E	H	J
T	D	E	S	K

For words across, Anna scores for *is, sue, suet, for, or, ram, am, ode,* and *desk*; for words down she scores for *hot, sold, old, urge, gee,* and *each*: a total score of 47, so she is the winner.

Also called Crosswords; Cross-Words; Five by Five; French Crosswords; Poker Crosswords; Scorewords; Squared Words; Wordsworth.

Compare Addiction; Jaxsquare; Scramble; Word Squares (2).

Crossword Puzzle *see* Crossword.

Crosswords *see* Crossword; Crossword Game.

Cryptarithms

⊕
②③ Any number of players   ✍ Played by writing
? Letters game   ✎ Played with pencils and paper

Object To create or solve sums in which letters represent numbers.

Procedure A cryptarithm is a puzzle in which letters are substituted for numbers in an arithmetical sum. Cryptarithms can be played as a game by trying to solve examples written by others, or by trying to create cryptarithms of one's own.

Example A typical cryptarithm looks like this:

$$\begin{array}{r} SEND \\ +MORE \\ +GOLD \\ \hline MONEY \end{array}$$

This can be solved if S = 5, E = 4, N = 7, D = 8, M = 1, O = 6, R = 2, G = 9, and L = 3:

$$\begin{array}{r} 5478 \\ 1624 \\ 9638 \\ \hline 16\,740 \end{array}$$

Cryptarithms can involve multiplication and division, as well as addition and subtraction, as in this example:

$$\begin{array}{r} NINE \\ \times FOUR \\ +FIVE \\ \hline FORTY\text{-}ONE \end{array}$$

The solution is that N = 9, I = 8, E = 5, F = 3, O = 0, U = 7, R = 4, and V = 6:

$$
\begin{array}{r}
9895 \\
\times 3074 \\
+3865 \\
\hline
30421\text{-}095
\end{array}
$$

Also called Alphametics.

Curtailments *see* Beheadments.

Cutlets *see* Bullets.

D

Daffy Definitions

② Any number of players	✍ Played by writing or speaking
? Punning game	✎ Played with pencils and paper, or with no equipment

Object To make humorous definitions.

Procedure Players choose (or are given) words for which they have to make humorous definitions. Often the definitions will use a pun (as in '*boomerang*: what you say to frighten a meringue' or '*carrion*: British comedy films') but sometimes they will be witty (as in '*alone*: in bad company', or '*cannibal*: someone who goes to a restaurant and orders the waiter').

Also called Daffynitions; Daft Definitions; Fractured Definitions.

Daffynitions *see* Daffy Definitions.

Daft Definitions *see* Daffy Definitions.

Decapitations *see* Beheadments.

Definitions

② Two or more players	✍ Played by speaking
? Guessing game	✎ Played with a dictionary

Object To guess words from their definitions.

Procedure One player takes the dictionary, chooses a word in it, and reads out the definition to the other players. They try to guess the word from its definition. The first person to guess correctly takes the dictionary and chooses the next mystery word.

Example Anna opens the *Concise Oxford Dictionary* and reads out this definition: 'A strong fabric sheet connected by springs to a horizontal frame, used by gymnasts etc. for somersaults.' Chas guesses correctly that the word is *trampoline*, so he takes the dictionary and reads out this definition: 'Living both on land and in water.' Tony guesses *seafaring* but Kate correctly guesses *amphibious*. She reads out the next definition: '500 sheets of paper.' Chas guesses *quire* but Anna correctly guesses *ream*.

Deflation

⓪②③ Any number of players	✍️ Played by writing or speaking
? Challenge game; punning game	✏️ Played with pencils and paper, or with no equipment

Object To find humorous ways of making economies in titles, etc.

Procedure Players imagine a situation in which attempts need to be made to economize in making films, publishing books, etc. They devise new titles for books, films, television programmes, etc. which reduce the resources originally needed. Thus *A Tale of Two Cities* might become *A Tale of One City*, *Ivan the Terrible* could be *Ivan the Not Very Nice*, and *An American in Paris* might save money by being retitled *An American in Scunthorpe*. This process can result in such titles as *The Two Musketeers*, *Snow White and the Three Dwarfs*, *None Flew Over the Cuckoo's Nest*, and *Desert Island Disc*.

Also called Wuthering Hillocks.

Compare Inflated Rhetoric.

Diamonds *see* Word Squares (1).

Dictionary *see* Acrostics (2); Dictionary Game; Words within Words.

Dictionary Definitions Game *see* Dictionary Game.

Dictionary Game

👥 Two or more players

❓ Guessing game

✍️ Played by writing and speaking

✎ Played with pencils and paper, and a dictionary

Object To guess the true meaning of chosen words.

Procedure One player chooses a fairly obscure word from a dictionary, and spells it out to the other players. All the players then write definitions of that word, preferably making them sound like genuine dictionary definitions. The first player reads out all the definitions, including among them the true definition. Each player then nominates which definition they think is correct. A point is awarded to everyone who chooses the correct definition and also to anyone whose false definition was accepted as true by another player.

Example Tony says: 'The word I have chosen from the *New Shorter Oxford English Dictionary* is *bubble-bow*.' The players write down their definitions and hand them to Tony, who reads them out, adding the correct definition from the dictionary. They are:

(1) A blob of dribble from a baby's mouth.
(2) A kind of tropical fish.
(3) A case in which ladies used to keep tweezers.
(4) A vain person; a show-off.
(5) An inflamed swelling on the elbow.

The players vote for the definition they believe is right. Then Tony tells them that the third sense is the correct one.

Also called Call My Bluff; Dictionary; Dictionary Definitions Game; Fictionary Dictionary.

Compare Encyclopaedia Fictannica; Side by Side.

Digrams

② Two or more players

? Word-building game

✍ Played by writing

✎ Played with pencils and paper

Object To think of words containing a given pair of letters

Procedure One player calls out two letters of the alphabet, and all the players have to write down (in a set time) all the words they can think of that contain those two letters together. The winner is the player who thinks of the largest number of words. Alternatively, players score one word for every word they make but two points for words that nobody else has written down. If desired, it can be ruled that the pair of letters must not start or end words.

Example Anna calls out the letters D and L. After five minutes, Kate has written down: *raddle, saddle, muddle, fiddle, diddle, oddly, sadly, badly, madly, cuddly, cordless, grandly, handle, handling,* and *swindle.* Other players have found some of these words, as well as *loudly, madly, wildly, endless, groundless, codling, wizardly, friendly, proudly, handler, handlebar, handlist, kindle, kindly, fondly, fondle,* and *blood-lust* (which will only score if hyphenated words are allowed).

Compare Trigrams.

Dingbats *see* Rebus.

Background Dingbats is a registered Trade Mark of Paul Sellers. The Dingbats boxed game is made by Waddingtons Games Limited under licence from Paul Sellers.

Dismissals

② Any number of players

? Punning game

✍ Played by writing or speaking

✎ Played with pencils and paper, or with no equipment

Object To think of appropriate punning ways to get rid of people.

Procedure The idea of this game is to think of a job, trade, or profession, and then to match it with a suitable punning verb to describe how people of this type can be dismissed or deprived of their normal function.

Examples Chiropodists can be defeated; wine merchants can be deported; musicians can be disbanded, disconcerted, or untuned; and Superman can be dismantled.

Also called Banishments.

Background Invented by Laurence Urdang.

Donkey *see* Ghosts.

Do This, Do That *see* Simon Says.

Double Acrostics

① ②③	Any number of players	✍	Played by writing
?	Word-finding game	✎	Played with a previously prepared puzzle, plus pencils and paper

Object To solve clues leading to words whose first and last letters spell words.

Procedure A double acrostic is like a simple acrostic—see Acrostics (1)—but the clues lead to words whose first *and* last letters spell out two different words. The clues are sometimes given in rhyme.

Example These clues lead to four four-letter words. The first and last letters of the words together form a phrase that seldom occurs (the second word of the phrase has its letters in reverse order).

(1) This can be runner or broad.
(2) This is a game on a board.
(3) Unbutton it.
(4) It's hard to sit.

Solution

B E A N

L U D O

U N D O

E X A M

(The first and last letters spell *blue moon*.)

Double-Bag Scrabble *see* Scrabble.

Double Centurion *see* Centurion.

Double-crostic

One player	Played by writing
? Grid game	✎ Played with a previously prepared puzzle

Object To guess words from clues and then transfer the letters into a grid to make a quotation, etc.

Procedure The player has to solve a number of clues which lead to words. Each letter of the words is numbered, so that the letters can be transferred to the numbered squares in a grid and thus make other words to build up a quotation, saying, etc. The squares in the grid are marked with letters to show which clue they refer to. The initial letters of the words also, if read vertically, spell the name of the quotation's author and sometimes the work from which the quotation is taken.

Example (The first clue has been filled in to give players a start.)

	Clues			Words		

A. He owes you in an old-fashioned way:

$\underline{\text{O}}$ $\underline{\text{W}}$ $\underline{\text{E}}$ $\underline{\text{T}}$ $\underline{\text{H}}$
36   9   18   28   20

B. Don't flog yourself silly in this game:

‾ ‾ ‾ ‾
1   33   42   22

C. A carpenter's wooden peg:

‾ ‾ ‾ ‾ ‾
27   13   46   30   43

D. Score less than a century:

‾ ‾ ‾ ‾ ‾ ‾
41   10   35   6   40   24

E. Assented — or dozed?:

‾ ‾ ‾ ‾ ‾ ‾
31   39   3   17   21   12

F. Quick and clever:

‾ ‾ ‾ ‾ ‾
26   7   32   19   48

G. Nearly:

‾ ‾ ‾ ‾ ‾ ‾
16   23   14   2   8   37

H. What unwise punters lose:

‾ ‾ ‾ ‾ ‾ ‾
45   29   4   34   38   11

I. You are only this:

‾ ‾ ‾ ‾ ‾
47   44   15   25   5

1 B	2 G	3 E		4 H	5 I		6 D	7 F	8 G
9 A W	10 D	11 H	12 E	13 C	14 G		15 I	16 G	17 E	18 A E
	19 F	20 A H	21 E			22 B	23 G	24 U	
25 I	26 F	27 C		28 A T	29 H	30 D	31 E		
	32 F	33 B	34 H	35 D	36 A O	37 G		38 H	39 E
40 D	41 D	42 B	43 C		44 I	45 H		46 C	47 I	48 F

> God in His wisdom made the fly
> And then forgot to tell us why.
> Ogden Nash, *The Fly*

Also called Leadergram.

Double Jeopardy

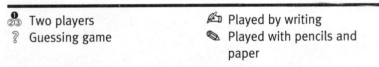

- Two players
- Guessing game
- Played by writing
- Played with pencils and paper

Object To guess a five-letter word chosen by one's opponent.

Procedure This is a variation of the game Jotto. It is played in the same way, but when a player suggests a 'test' word, that player must say how many letters it has in common with his or her own secret word. This makes the game more challenging, as players must think of test words that gain information from their opponent while giving away as little information as possible about their own word.

Double Meanings *see* Homonyms.

Doublets

⚉ Any number of players   ✍ Played by writing
? Challenge game   ✎ Played with pencils and paper

Object To change one word into another in stages, each stage making a new word.

Procedure Players are given, or choose, one word to change into another which has the same number of letters. The two words should preferably be related in their senses: like *dog* and *cat*, or *love* and *hate*. Players change one letter at a time and, at each stage, a new word has to be made. The winner is the person who completes the change in the smallest number of moves.

Example The writer Lewis Carroll invented this game, and one of his earliest examples was changing *head* into *tail*, which he did thus:

> HEAD
>
> HEAL
>
> TEAL
>
> TELL
>
> TALL
>
> TAIL

Carroll also allowed players to rearrange the letters instead of changing a letter, as in this example:

> IRON
>
> ICON
>
> COIN
>
> CORN
>
> CORD
>
> LORD
>
> LOAD
>
> LEAD

Also called Laddergrams; Laddergraphs; Letter by Letter; Letter Change; Passes; Stepwords; Transformations; Transitions; Transmutations; Word Alchemy; Word Chains; Word Golf; Word Ladders; Word Links; Word Ping-Pong.

Compare Triplets.

Droodles *see* Rebus.

Dumb Crambo

2/3 Two or more players

? Active game; guessing game; rhyming game

✍ Played by miming

✎ No equipment needed

Object To guess a word by means of rhyming words.

Procedure The players divide into two teams. One team goes outside the room, while the other team chooses a word—preferably a verb that rhymes with several other words. The players outside the room are told a word that rhymes with the mystery word. They try to guess between themselves what the answer may be, and they re-enter the room to act out this answer in mime. If they have guessed the word wrongly, the other team hisses or boos them, and they leave the room to make another guess. When they mime the correct answer, the other team applauds them, and leaves the room to become the guessers. Alternatively, the game can be spoken instead of mimed (see Crambo).

Example One member of the team inside the room tells those outside that the mystery word rhymes with *hand*. The outside team comes in and mimes *stand* (by sitting down and then standing) but is booed. They then decide to mime *band*, pretending to play musical instruments, but again they are booed. For their third attempt, they mime *land* (by pretending to be aircraft) and are applauded for having guessed correctly.

Compare Botticelli; Charades; Crambo; The Game.

E

Earth, Air, Fire, and Water *see* Earth, Air, Water.

Earth, Air, Water

 Four or more players
? Challenge game

 Played by speaking
Played with a small object such as a handkerchief or ball

Object To think of words that fit the categories *Earth, Air,* or *Water*.

Procedure The players stand or sit in a circle. One player throws a small object (such as a handkerchief, ball, etc.) to one of the other players, shouts either 'Earth', 'Air', or 'Water', and starts counting up to ten. Before the number ten is reached, the other player must name a creature that lives in the named region. For instance, if the category is 'Earth', possible answers could be *tiger* or *slug*. If the category is 'Air', possible answers could include *swallow* or *wasp*. If a player thinks of a suitable creature in time, that player becomes the person to throw the object at someone else but, if the player cannot think of a suitable answer, the person who counted to ten has a second go, throwing the object to another player. In a variation of the game called *Earth, Air, Fire, and Water*, the player has to remain silent if 'Fire' is shouted.

Also called Fire, Air, and Water.

Encyclopaedia Fictannica

⟨2⟩⟨3⟩ Two or more players
❓ Guessing game

✍ Played by writing
✎ Played with pencils and paper, and an encyclopedia

Object To guess a genuine entry from an encyclopedia.

Procedure This game is similar to the Dictionary Game except that it uses an encyclopaedia instead of a dictionary. One player chooses a fairly obscure entry from an encylopedia, and gives its title to the other players. All the players then write entries for that title, preferably making them sound like genuine entries from an encyclopedia. The first player reads out all the entries, including among them the genuine entry. Each player then nominates which entry they think is correct. A point is awarded to everyone who chooses the correct entry and also to anyone whose false version is accepted as true by another player.

Example Kate says: 'The entry I have chosen from the Encyclopedia is *Chibchas*.' The players write down their entries and hand them to Kate, who reads them out, adding the correct version from the encyclopedia. They are:

(1) A species of extinct mammoth-like creatures believed to have inhabited the islands around Indonesia in prehistoric times.

(2) South American Indians of Colombia, whose civilization was overthrown by the Spaniards in 1538. They used to powder their chief with gold dust, a rite which encouraged the legend of El Dorado.

(3) A language spoken in Kenya and Uganda. It is allied to the Tanzanian *Otombe* and the *N'Zuta* dialect of the Zambezi region.

(4) A suburb of Calcutta in India, noted for its confectionaries.

(5) In Greek mythology: the gods that dwell in the underworld, guarded by Charon and the dog Cerberus.

The players vote for the version they believe is correct. Then Kate tells them that the second entry is the correct one.

Compare Dictionary Game.

Endings

👥 Two or more players   ✍ Played by writing or speaking

❓ Challenge game   ✎ Played with pencils and paper, or with no equipment

Object To think of words with particular endings.

Procedure This game can be played in two ways. Players can take turns to ask other players to think of a word that ends with a particular series of letters. Alternatively, players can be given a list of word-endings and challenged to write down one word for each ending. The winner is the first player to write down suitable words for the whole list.

Example

TONY: Can you think of a word that ends with –RY?

KATE: *Try.*

ANNA: *Inquiry.*

CHAS: *Descry.* Can you think of a word that ends with –YER?

TONY: *Player.*

KATE: *Soothsayer.*

ANNA: *Lawyer.* Can you give me a word that ends with –CEFUL?

CHAS: *Resourceful.*

TONY: *Peaceful.*

KATE: *Disgraceful.* Can you think of a word that ends with –UIA?

ANNA: *Alleluia.*

CHAS: *Colloquia.*

TONY: I can't think of one. (*He has to drop out of the game.*)

Also called Word Endings.

Enigma

②ⓓ Any number of players
⁇ Guessing game; poetic game

✎ Played by writing
✍ Played with pencils and paper

Object To write or solve a complicated riddle

Procedure An enigma is a riddle, quite often in verse and usually quite long. Enigmas differ from riddles either in being written in poetic form or by being much more complex than simple riddles. They are descriptions of something, ingeniously worded in a misleading way which makes them difficult to unravel.

Example

> What's that which all love more than life,
> Fear more than death or mortal strife?
> That which contented men desire,
> The poor possess, the rich require,
> The miser spends, the spendthrift saves,
> And all men carry to their graves?

Solution Nothing.

Equivocal Verse *see* Equivoque.

Equivoque

②ⓓ Any number of players
⁇ Challenge game; poetic game

✎ Played by writing
✍ Played with pencils and paper

Object To write a poem which has two different meanings.

Procedure The aim of an equivoque is to write a poem which can be read in two ways, so as to give two completely different meanings. This is usually achieved by arranging the lines so that, if only the first half (or second half) of each line is read, it gives a different sense from the whole poem. The difference in meanings can sometimes be achieved by reading alternate lines of the poem, or by changing its punctuation.

Example The whole poem below reads like a criticism of marriage, but reading the first or second column on its own gives an entirely different picture:

The man must lead a happy life	Who's free from matrimonial chains
Who is directed by his wife	Is sure to suffer for his pains
Adam could find no solid peace	When Eve was given for a mate
Until he saw a woman's face	Adam was in a happy state.

Also called Equivocal Verse.

Explain That *see* Stepping Stones.

F

Famous Last Words

Any number of players	✒ Played by writing or speaking
❓ Challenge game	✎ Played with pencils and paper, or with no equipment

Object To think of what celebrities might say just before they die.

Procedure Many historical characters have uttered famous last words, just before they died. For instance, Anne Boleyn is supposed to have said, before her head was chopped off: 'The executioner is, I believe, very expert, and my neck is very slender.' The subject of 'famous last words' can be used as a game, where players try to think of suitable 'last words' for famous people or for animals, perhaps suggested by other players. For example, a famous cricketer might say; 'Out? I don't believe it!' Socrates might have said: 'Thank you. I needed a drink.' And the last words of a cat might be: 'I am going to look at a king.'

Fictionary Dictionary *see* Dictionary Game.

Fill-Ins

Two or more players	✒ Played by writing
❓ Guessing game; word-finding game	✎ Played with pencils and paper

Object To think of words that start and end with particular letters.

Procedure This game is similar to Tops and Tails, except that players have to find the missing letters in a pre-arranged list which includes only the first and last letters of each word. One player prepares duplicated lists of first and last letters for words that all contain the same number of letters (say, five or six letters). The other players then have to fill in the missing letters to make words. The winner is the player who completes the list first, or the player who completes the most words within a given time-limit.

Example Chas prepares a list of the first and last letters of ten six-letter words:

A — L
I — D
S — T
C — Y
W — N
F — L
T — E
L — M
G — O
Z — H

In two minutes, Kate writes down these words: *actual, imbued, submit, comely, weapon, finial, tinkle, lissom, gaucho,* and *zenith.* Tony writes down: *animal, intend, strait, comedy, within, funnel, timbre,* and *logjam.* He cannot think of words for G–O or Z–H, so he scores only eight points, and Kate wins with ten.

Compare Tops and Tails.

Find the Word

🎯 Two or more players		🖊 Played by writing and speaking	
❓ Word-finding game		🖊 Played with pencils and paper	

Object To guess a word by solving clues to groups of its letters.

Procedure One player writes down a long word and numbers each letter. This player tells the other players how many letters the word contains, and they each write down that number of dashes and number them. The first player then gives clues to words that can be made from groups of letters chosen from the secret word, giving the numbers of the letters chosen. The other players write down the answers in the appropriate numbered spaces, trying to build up the word until they have enough letters to guess it correctly. The person who first guesses the word correctly becomes the one who chooses the next mystery word.

Example Kate tells the other players that her word contains 11 letters, and they write down 11 numbered dashes. Kate then tells them that letters 3, 7, and 6 make a friend. Those who can guess what she means write down the letters P, A, and L (the word *pal*) in the places numbered 3, 7, and 6, so that their papers look like this:

$$\underset{1}{\_}\ \underset{2}{\_}\ \underset{3}{P}\ \underset{4}{\_}\ \underset{5}{\_}\ \underset{6}{L}\ \underset{7}{A}\ \underset{8}{\_}\ \underset{9}{\_}\ \underset{10}{\_}\ \underset{11}{\_}$$

She then tells them that letters 8, 5, 2, and 11 are 'quite correct'. The clever players write down the letters T, R, U, and E in those spaces. Their papers then look like this:

$$\underset{1}{\_}\ \underset{2}{U}\ \underset{3}{P}\ \underset{4}{\_}\ \underset{5}{R}\ \underset{6}{L}\ \underset{7}{A}\ \underset{8}{T}\ \underset{9}{\_}\ \underset{10}{\_}\ \underset{11}{E}$$

Anna guesses that the word is *superlative*, which is correct, and so she chooses the next mystery word.

Fire, Air, and Water *see* Earth, Air, Water.

Five by Five *see* Crossword Game.

Fizz *see* Buzz.

Fizz-Buzz *see* Buzz.

Follow On

- Two or more players
- Challenge game; cumulative game
- Played by speaking
- No equipment needed

Object To think of a phrase that starts with the second element of a preceding phrase.

Procedure The first player calls out a word or phrase which consists of two parts, like *junk shop*. The next player has to say a word or phrase that starts with *shop*—for example, *shop-window*. And so on, round the players in turn. Any player who cannot think of a suitable word or phrase is out of the game, and the winner is the last player left. Anyone can challenge another player's word, which is unacceptable if it is not included in the dictionary. The game can also be played with the last syllable of one word forming the start of the next word, so that *hospital* might lead to *talon, lonely, lycanthrope*, and so on. For young children, the last *letter* of one word can be the start of the next word, so that the chain might be: *mouse, ever, rebel, lock*, etc.

Example

ANNA: *Shoe-tree.*
KATE: *Tree-trunk.*
CHAS: *Trunk call.*
TONY: *Call off.*
ANNA: *Offside.*
KATE: *Sideways.*
CHAS: *Wayside.*
ANNA: That's not acceptable. *Wayside* is a combination of *way* and *side*, not *ways* and *ide*! (*Chas is out of the game.*)
TONY: *Sidelong.*
ANNA: *Long-winded.*

(Kate cannot think of a phrase starting with *winded*, but nor can anybody else, so the four players start another round of the game.)

Also called Trailers; Word Chain.

Compare Head to Tail.

Forbidden Words *see* Yes and No (1).

Fore and Aft *see* Alpha.

Forehead Game *see* Foreheads.

Foreheads

⊕ Three or more players	✎ Played by writing and speaking
? Guessing game	✎ Played with sticky labels and pencils

Object To guess a word which is written on a label stuck to one's forehead.

Procedure Each player writes a four-letter word on a sticky label, and sticks it on the forehead of the player on his or her right. Every player can see all the words except the one on his or her own forehead. Each player then, in turn, says a four-letter word made up of letters which are on other people's foreheads. The chosen word can include the same letter more than once. Players can gradually build up an idea of the words on their own foreheads by noting which letters are mentioned that do *not* occur in any of the words on other people's foreheads. Players can use their 'go' to guess their own word. If the guess is correct, that player wins; if the guess is incorrect, that player stays in the game but cannot make any more guesses.

Example Chas sticks the word *fist* on Anna's forehead; Anna sticks *goes* on Tony's forehead; Tony sticks *able* on Kate's forehead; and Kate sticks *call* on Chas's forehead.

CHAS: *Ball.* (This gives no help to Anna or Tony, as they can see all these letters on other people's foreheads, but it tells Kate that her word contains the letter *b*, as she cannot see this letter on anyone else's forehead.)

ANNA: *Sago.* (This is no help to Kate or Chas, but it tells Tony that his word contains the letters *g* and *o*.)

TONY: *Cast.* (This tells Anna that her word contains a *t*.)

KATE: *Soft.* (This tells Anna that her word also includes an *f*.)

CHAS: *Bits.* (This tells Anna that her word also contains an *i*.)

ANNA: Is my word *fist*?

EVERYONE ELSE: Yes!

Also called Forehead Game.

Background Invented by Dave Silverman.

Fortunately, Unfortunately *see* Good News, Bad News.

Found Poems *see* Prose Poems.

Fractured Definitions *see* Daffy Definitions.

Fractured French

ⓐ Any number of players	✍ Played by writing or speaking
? Punning game	✎ Played with pencils and paper, or with no equipment

Object To create punning definitions of French phrases.

Procedure The aim is to take a French word or phrase and give it a comical definition, usually by using a pun. This is often achieved by imagining how the phrase might be pronounced in English, as in *pas du tout* which is translated as 'father of twins' (because *pas du tout* sounds like *pa de two*). The game can also be played with words and phrases from other languages.

Examples

Cul-de-sac: sack of coal.
À la carte: on the wagon.
Bidet: two days before D-Day.
Mise en scène: there are mice in the river.
Autobahn: garage.

Bambino: a reprimand by a mother deer.

Honi soit qui mal y pense: I honestly think I am going to be sick.

Background The game gets its name from F. S. Pearson's book *Fractured French* (1951).

Free Association *see* Associations.

French Crosswords *see* Crossword Game.

Fusions

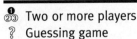

⚇ Two or more players	✍ Played by speaking
❓ Guessing game	✎ No equipment needed

Object To guess words that consist of two or more other words.

Procedure This game is based upon the fact that many words can be divided up so as to make two or more other words. Thus *fortune* can be divided into *for* and *tune*, while *together* can be divided into *to*, *get*, and *her*. Players look for words like this and ask the other players to guess the whole words from clues.

Example

TONY: Can you add a word for an automobile to a word for a
       domesticated animal to make a floor-covering?

ANNA: *Car-pet*. Can you add a kind of snake to a word for 'anger' to get
       a word that means 'to have ambitions'?

CHAS: *Asp-ire*. Can you add a containing word to a kind of tree and a
       woman's name to make a hospital?

KATE: *In-fir-mary*.

Compare Word Divisions.

G

Gallows *see* Hangman.

The Game

⊕ Two or more players (usually two teams)	✍ Played by miming
? Active game; guessing game	✎ No equipment needed

Object To guess a word or phrase that is acted out in mime, usually in syllables.

Procedure Those present divide into two teams and a referee. The referee prepares a list of words or phrases, often titles of films, television programmes, plays, or books. The referee shows the first word or phrase to the first player from each team. These players then have to mime the word to their team. Whoever guesses it, goes to the referee to tell the referee and be given the next word or phrase to be mimed. The winning team is the one that guesses the most items in a fixed time, or reaches the end of the referee's list first.

There are several alternative versions of The Game. In one version, one member of each team thinks of a word or phrase which they then try to mime to their teams. The Game can also be played without teams, one person simply trying to convey a chosen thing to everyone present. Whoever guesses it becomes the next mimer.

In the form popularized on television as *Give Us a Clue*, one member of a team at a time is given a word or phrase by the question-master, to mime within two minutes. If the mimer's team fails to guess it, the other team can win points by guessing it. The television version has established ways of indicating whether a phrase is a song title (by miming the action of singing), a film (by pretending to operate a film camera), a play (by miming the opening of curtains), or a television programme (by drawing a rectangle in the air). Other conventions for the mimer include raising a number of fingers to indicate how many words the phrase contains, and laying one's fingers across one's arm to indicate the number of syllables.

Example Kate decides to mime the song title 'All the Things You Are'. She starts by opening her mouth wide and throwing out her arms. The guessers agree that this must be the title of a song. Kate holds up five fingers, to show that the title contains five words. Then she holds up one finger to show that she is going to mime the first word. She tries to mime the word *all* by making huge sweeping gestures with her arms. Anna guesses *aeroplane*; Chas suggests *aerobics*; and Tony—thinking he has the answer—shouts 'Windmills of Your Mind'! Kate shakes her head and does some more mime, eliciting the guesses *exercises*, *everything*, and eventually *all*.

She holds up two fingers to show that she will now mime the second word, which she does by placing one finger across the top of another: the conventional sign for *the*. She then mimes *things* by pretending to be 'The Thing from Another Planet'. Guesses include *King Kong, Godzilla, Tony*, and finally *thing*. Kate makes a gesture drawing her hands apart, indicating that the word should be lengthened. Tony, realizing it is *things*, guesses 'All the Things You Are'—and he becomes the next person to do the miming.

Also called Give us a Clue; Pantomime Race.

Compare Charades; Crambo; Dumb Crambo; Pictures.

Game of the Name *see* Awful Authors.

Garibaldi *see* Botticelli.

Geography

Two or more players	Played by speaking
Word-finding game	No equipment needed

Object To think of words that start with the last letter of a preceding word.

Procedure The players choose a category: for example, countries, cities, trees, film titles, or famous writers. The first player calls out a word that fits the chosen category. The second player then has to say another word in that category that starts with the last letter of the previous word. Play continues around the players in turn. Players drop out if they cannot think of a suitable word within five seconds. No word can be repeated that has already been used.

Example The players choose the category 'Composers'.

ANNA: Beethoven.
KATE: Nicolai.
CHAS: Ibert.
TONY: Tippett.
ANNA: Tallis.
KATE: Stanford.
CHAS: Delius.
TONY: Schubert.
ANNA: Tchaikovsky.

(None of the players can think of a composer starting with Y, so they begin another round.)

Also called Grab On Behind; Heads and Tails; Last and First; Last Shall Be First; Trailing Cities.

Get the Message

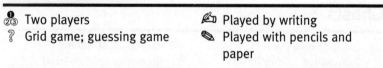

| 👥 Two players | 🖎 Played by writing |
| ❓ Grid game; guessing game | ✎ Played with pencils and paper |

Object To guess a phrase which your opponent has hidden in a grid.

Procedure This game is similar to Quizl, except that a *phrase* (not a word) is hidden in a grid by arranging the letters in a sequence that can be traced by going right or left, up or down, from one square to another. Thus the phrase 'Power to the people' might be inserted in the grid like this, starting in the top right-hand corner:

	1	2	3	4	5	
	L	P	O		P	6
	E	P	E	W◄─O		7
		E		E─►R		8
		H	T◄─O◄─T			9
						0

Players in turn choose a numbered square and their opponent reveals the letter which occupies that square—or says that the square is blank. Instead of choosing a square, a player can make a guess at the whole phrase. The winner is the first person to guess the other player's phrase.

Also called Phrase Maze.

Compare Quizl; Word Battleships.

Background Invented by David Parlett.

Ghost *see* Ghosts.

Ghosts

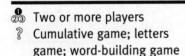

Two or more players	Played by speaking
Cumulative game; letters game; word-building game	No equipment needed

Object To build up letters without completing a word.

Procedure The first player thinks of a word of three or more letters, and calls out its first letter. The second player adds another letter which continues but does not complete a word, and so on, until one player is forced to finish a word. Any player who adds a letter can be challenged by the next player to say what the word will be. Any player who loses such a challenge, or completes a word, becomes 'a third of a ghost'. When this player loses again, he or she becomes 'two-thirds of a ghost' and the third time becomes 'a whole ghost' and has to drop out of the game.

Example Anna thinks of the word *jet* and says J. Chas thinks of *jester* and says E. Kate thinks of *jettison* but cannot say T because that would make a whole word. So she thinks instead of *jelly* and says L. Tony has to say another L, because he cannot think of any word except for *jell* or *jelly*. He thus completes a word and he becomes 'one-third of a ghost'.

Also called Chain Spelling; Donkey; Ghost; Monkey; Wraiths.

Compare Anaghost; Superghosts.

Giotto *see* Jotto.

Give Us a Clue *see* The Game.

Goldwynisms *see* Irish Bulls.

Good News, Bad News

⓪②③ Two or more players	🖾 Played by speaking
❓ Cumulative game	🖋 No equipment needed

Object To build up a story consisting of alternately good and bad pieces of news.

Procedure The first player starts by stating a piece of good news, such as 'The good news is: we are all going on a seaside holiday.' The next player then has to state a piece of bad news which tends to contradict the good news, such as 'The bad news is: it's a seaside holiday in the Antarctic.' And so on round the players, with each player adding alternate pieces of good or bad news in turn.

Example

KATE: The good news is that we are all going to get an extra helping of pudding.

ANNA: The bad news is that it's semolina.

CHAS: The good news is that each portion has a large dollop of jam on top.

TONY: The bad news is that there's a dead fly in each dollop of jam . . .

Also called Fortunately, Unfortunately.

Grab On Behind *see* Geography.

Grandmother's Trunk

👥 Two or more players	✍ Played by speaking
❓ Alphabetical game; cumulative game	🖐 No equipment needed

Object To remember an ever-growing alphabetical list.

Procedure The first player says 'My grandmother keeps an A_____ in her trunk' (giving a word starting with the letter A). The second player has to add an article beginning with B, and so on, listing all the objects so far mentioned. Anyone who cannot remember the whole list has to drop out of the game.

Example

KATE: My grandmother keeps an *anorak* in her trunk.

CHAS: My grandmother keeps an *anorak* and a *balloon* in her trunk.

TONY: My grandmother keeps an *anorak*, a *balloon*, and a *constable* in her trunk.

ANNA: My grandmother keeps an *anorak*, a *balloon*, a *constable*, and a *Dobermann* in her trunk.

KATE: My grandmother keeps an *anorak*, a *balloon*, a *constable*, a . . . er . . .'

and is out of the game.

Also called Portmanteau.

Compare Alphabet Dinner; Hypochondriac; I Packed My Bag; I Went to Market; Traveller's Alphabet.

Group Limericks

 Two or more players
? Cumulative game; poetic game

🖎 Played by speaking
✎ No equipment needed

Object To make up a limerick, with players each contributing a line.

Procedure The first player makes up the first line of a limerick, the second player adds the second line, and so on round the players. The lines must rhyme in the proper style of a limerick (A–A–B–B–A). Any player who cannot think of a suitable line is out of the game.

Example

CHAS: There was an old woman from Gloucester . . .

ANNA: Whose husband thought he had lost her . . .

KATE: When he looked in the freezer . . .

TONY: He found the old geezer . . .

CHAS: But he didn't know how to defrost her!

Also called Limericks; Limericksaw.

Group Story *see* Rigmarole.

Guessing Proverbs *see* Proverbs.

Guggenheim

②③ Two or more players   ✍ Played by writing
? Word-finding game   ✎ Played with pencils and paper

Object To find words that begin with particular letters of the alphabet and that fit particular categories.

Procedure A word of about six letters is chosen, either at random or from the pages of a book, but it should contain no letter more than once. Players agree on a number of categories (animals, towns, trees, book titles, etc.) and write them down the left-hand side of their paper and the letters of the chosen word along the top. They then try, for each category, to write words that start with the letters of the chosen word. A time-limit is set, and a player scores points for each word that no other player has written (or, if preferred, one point for each of the players who has not written that word).

Example The four players each suggest a category, and they write them in a column down the left-hand side of their pieces of paper. Anna pulls a book from the shelf and Kate tells her to pick out the second word in the fifth line on page 67. The word is *pints*, so they write P I N T S along the top of their sheets. They then have five minutes to write down their answers. Tony's list looks like this:

	P	I	N	T	S
Animals	pig	ibis	newt	tiger	sheep
Countries	Peru	Ireland	Netherlands	Tanzania	Switzerland
Female names	Pulcinella	Ivy	Nerissa	Thelma	Susan
Film titles	Peter Pan	I Am a Camera	Never on Sunday	Tarzan of the Apes	Sex, Lies & Videotape

Tony scores one point each for *ibis*, *Netherlands*, *Pulcinella*, *Nerissa*, and all the film titles, which nobody else has written, so he scores nine points. In this game, you need to think of words which the other players are unlikely to write.

Also called Categories; Initial Letters; Knowledge List; One for Each; (as a boxed game) Scattergories.

Compare Outburst!

H

Hack Saws Resharpened *see* Perverbs.

Hanging the Man *see* Hangman.

Hangman

👥 Two or more players	✍ Played by writing
❓ Guessing game	✏ Played with pencils and paper

Object To guess a word, one letter at a time, before one uses up one's chances and is 'hanged'.

Procedure One of the two players thinks of a word or short phrase, and draws dashes on a piece of paper for each letter it contains. The other player then guesses letters, one at a time. If the guessed letter is in the word, the first player writes it in the appropriate space or spaces; if that letter is *not* in the word, the first player starts to draw a gallows, which is added to every time a wrong letter is guessed. If the word is not guessed before the gallows is completed (with a matchstick figure hanging from it), the guesser loses the round. If more than two people are playing, they guess letters in succession.

Example Kate tells Chas that the hidden word contains six letters, and she draws six dashes on her paper:

Chas suggests the letter E, but Kate says it is not in the hidden word, and she draws the base of the gallows. Chas guesses S, but this is not in the word either, so Kate adds an upright post to the base of the gallows. After three correct guesses and five incorrect ones, the word spaces look like this:

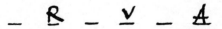

and the gallows looks like this:

The gallows is usually completed after 11 incorrect guesses, in this order:

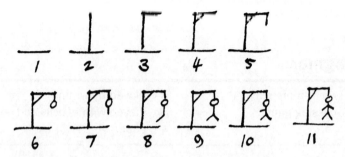

In some versions of the game, the gallows is built more quickly (omitting the base and the matchstick figure's arms) or more slowly (with more component parts). Players should agree beforehand on how the gallows is to be constructed.

Fortunately for Chas, he guesses that the word is *trivia* before the hanging is completed.

Also called Gallows; Hanging the Man; Probe.

Hankety Pankety *see* Stinky Pinky.

Hank Pank *see* Stinky Pinky.

Hanky Panky *see* Stinky Pinky.

Headline Game *see* Headlines.

Headliners *see* Headlines.

Headlines

😀 Two or more players   ✍ Played by writing or
                                      speaking

❓ Cumulative game   ✏ Played with pencils and
                                        paper, or with no equipment

Object To build up a newspaper headline.

Procedure The aim of this game is to build up a newspaper headline, with each player in turn contributing a word. The game can be spoken, with each player suggesting one word until a headline is completed. Alternatively, the game can be written, with each player writing a word on a piece of paper which is then folded over (in the style of Consequences) and passed to the next player. Players can be asked to supply words in the following five categories: (1) an adjective or adjectival phrase; (2) a noun (e.g. a kind of person or animal); (3) a verb; (4) the object of the verb; and (5) where the action occurred. This may result in a ridiculous headline such as: INTOXICATED DOCTOR FINDS CAT ON THE MOON.

There are two other forms of headline games. One requires players to make up headlines describing historical or other events in an indirect way. The other players have to guess which event is being headlined. So the headline FLAMING PUDDING STARTS CITY TRAGEDY could describe the Fire of London starting in Pudding Lane, while KING'S SON KILLS UNCLE AND THEN DIES might refer to Shakespeare's play *Hamlet*. Another game simply consists of trying to think of comical or nonsensical headlines, such as: PIANIST GETS ORGAN TRANSPLANT or BIG BOOM IN DYNAMITE SALES.

Also called Headline Game; Headliners.

Heads and Tails   *see* Bullets; Geography.

Heads are Tails   *see* Alpha.

Head to Tail

 Any number of players

? Challenge game; cumulative game

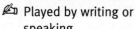 Played by writing

Played with pencils and paper

Object To make a string of words or phrases in which the second element of one becomes the first element of the next.

Procedure A combination of two words is chosen, and players then try to write the longest sequence of words in which the second part of one word is the first part of the next. If more than one person plays, the game can be made competitive, with the winner being the player who writes the longest list.

Example Anna and Tony chose to start with the combination *face-cream*. Anna continued the sequence with: *cream cheese, cheesecake, cakewalk, walkway, way-out, outsmart,* and *smart alec.* Tony's list was: *cream cracker, cracker-barrel,* and *barrel-chested,* but he could think of no phrase starting with *chested,* so Anna won, as her list was longer.

Compare Follow On.

Hermans

 Any number of players

? Punning game

 Played by writing or speaking

Played with pencils and paper, or with no equipment

Object To think of suitable names for people who say specified things.

Procedure This game resembles Tom Swifties, in which a punning adverb is matched to a statement. Hermans also use puns: by choosing a suitable name for a person who says a particular thing.

Examples

'She's my woman,' said Herman.
'The cat scratched me!' said Claude.
'These shoes make me wobble,' said Lucille.
'Does this skirt cover my legs?' said my niece.
'What comes after H?' asked I, Jay, Kay, Ella, and Emma.

Heteronyms

 Two or more players   Played by speaking
? Guessing game No equipment needed

Object To guess two words which are spelt the same but pronounced differently.

Procedure One player thinks of a pair of words which have the same spelling but different pronunciations and senses, such as *bow* (meaning the front of a ship) and *bow* (meaning bending your body as a mark of respect). This player then asks the other player or players to guess the words from their definitions. The player who guesses correctly can choose the next pair of mystery words.

Example

CHAS: Which two words are these, which are spelt the same but pronounced differently? One means 'to leave'; the other means 'a large expanse of sand'.

KATE: *Desert* (pronounced di-**zert**) and *desert* (pronounced **dezz**-ert). Can you guess this pair of heteronyms? One means 'a sick person'; the other means 'not acceptable'.

ANNA: *Invalid* (pronounced **in**-va-leed) and *invalid* (pronounced in-**val**-id). Try these two: one means 'to say no'; the other means 'rubbish'.

TONY: *Refuse* (pronounced ri-**fuse**) and *refuse* (pronounced **ref**-ewss).

Compare Homonyms; Homophones.

Hidden Proverbs *see* Proverbs.

Hidden Words (1)

②③ Any number of players	✍ Played by writing or speaking
? Word-finding game	✎ Played with pencils and paper, or with no equipment

Object To find words hidden inside sentences.

Procedure This game can be played in several ways. One player may hide a word inside a sentence for another player to find. For example, a player might be asked to find an animal hidden inside the sentence 'He made errors all the time.' The answer is *deer*, which is hidden in 'ma*de er*rors'. Alternatively, a fixed sentence such as a proverb can be suggested, and players have to find words inside it. Thus the proverb 'A stitch in time saves nine' includes the words *as, stitch, tit, titch, itch, chin, hint,* and several others. Another method of play is to propose single words that contain other words inside them. For instance, there are animals inside the words *doggerel, scatter, kneel, bath,* and *duration*.

Also called Buried Words; Concealed Words; Hide and Seek.

Hidden Words (2) *see* Words within Words.

Hide and Seek *see* Hidden Words (1).

Hinkety Pinkety *see* Stinky Pinky.

Hink Pink *see* Stinky Pinky.

Hinky Pinky *see* Stinky Pinky.

Hobbies

②③ Two or more players	✍ Played by speaking
? Challenge game; letters game	✎ No equipment needed

Object To think of phrases that match people's initials.

Procedure The first player thinks of the name of a celebrity, and uses it in a question to the second player, for example: 'What's your hobby, William Shakespeare?' The second player must think of a phrase which starts with the same initials as William Shakespeare: it could be 'Washing Socks' or 'Writing Sonnets'. And so on, round the players. Anyone who cannot think of a suitable phrase is out of the game, and the winner is the last player left. Alternatively, the game can be played with a question-master asking questions of each player in turn, or presenting the players with lists of celebrities for which to find appropriate phrases within a given time.

Example

ANNA: What's your hobby, Jane Austen?
TONY: Judging Athletics. What's your hobby, Mick Jagger?
CHAS: Making Jam. What's your hobby, Igor Stravinsky?
KATE: Investigating Sausages.

Also called Hobby-Horse; Leading Lights.

Compare Initial Letters (1).

Hobby-Horse *see* Hobbies.

Holorimes *see* Anguish Languish.

Homonyms

Object To guess two words which have the same spelling but different meanings.

Procedure One player thinks of a pair of words which are spelt and pronounced the same but have different meanings, such as *burn*—which can mean to destroy by fire and a stream in Scotland. This player then asks the other player or players to guess the words from these definitions. The player who guesses correctly chooses the next pair of mystery words.

Example

TONY: Can you guess these two homonyms? One means 'a friend'; the other is a position in the game of chess.

KATE: *Mate.* What about this pair of words? One is a period of relaxation; the other means 'the remainder'.

ANNA: *Rest.* Try these two: the first one means 'in good health'; the other is a place where you can get water.

CHAS: *Well.*

Also called Double Meanings.

Compare Heteronyms; Homophones; Triple Meanings.

Homophones

Object To guess two words which are pronounced the same but spelt differently.

Procedure One player thinks of a pair of words which sound the same but are spelt differently and have different meanings, such as *bare* and *bear*. He asks the other player or players to guess the words from definitions which he supplies. The player who guesses the pair of words correctly chooses the next pair of mystery words.

Example

KATE: Try to identify this pair of words. One means 'yourself' and the other means 'a female sheep'.

CHAS: *You* and *ewe*. What are these two homophones? One means 'the atmosphere'; the other means 'an inheritor'.

ANNA: *Air* and *heir*. Can you guess this pair? One is a ring of bells; the other is the outside of a fruit.

KATE: *Peal* and *peel*. What are these two? One is a number; the other is a river.

ANNA: I've got it! *Seven* and *Severn*.

Compare Heteronyms; Homonyms.

Howlers *see* Malapropisms.

How, When, and Where

🔢 Three or more players	✍️ Played by speaking
❓ Guessing game	✎ No equipment required

Object To guess a word from answers given to three questions.

Procedure One person leaves the room and the remainder choose a word, preferably a noun that has several meanings (like *chip*, *skate*, or *tin*). The player outside the room then returns and puts three questions to each of the other players in turn: '*How* do you like it?', '*When* do you like it?', and '*Where* do you like it?' When all the players have given their answers, the guesser suggests what the word is. If the guesser is wrong, he or she goes out of the room again. If the guesser is correct, he or she indicates who gave the answer that chiefly suggested the word, and *that* person is the next one to leave the room. In a variation of the game called News Reporter, the guesser asks: 'When/Where/How/Why do you use it?'

Example Tony goes out of the room, and the others choose the word *paper*. Tony re-enters.

TONY (*to Anna*): How do you like it?
ANNA: White.
TONY: When do you like it?
ANNA: When I'm writing.
TONY: Where do you like it?
ANNA: On the table.
TONY (*to Chas*): How do you like it?
CHAS: Full of information.
TONY: When do you like it?
CHAS: On Sunday morning.
TONY: Where do you like it?
CHAS: In bed.
TONY (*to Kate*): How do you like it?
KATE: Coloured.
TONY: When do you like it?
KATE: At Christmas.
TONY: Where do you like it?
KATE: Round Christmas presents.

Tony correctly guesses that the word is *paper*, and says that Kate's last answer was what gave it away, so Kate goes out for the next round.

Also called How, When, Where; How, Where, When; Where, When, and How; Why Do You Like It?

How, When, Where *see* How, When, and Where.

How, Where, When *see* How, When, and Where.

Huntergrams

👥 Any number of players	✍ Played by writing
❓ Anagrams game	✏ Played with pencils and paper

Object To solve anagrams in a poem.

Procedure Huntergrams are pieces of verse in which blanks represent words which are anagrams of one another. The aim is either to solve ready-made puzzles of this sort, or to write them for your own amusement.

Example

> 'These —— seem awfully large,' I said
> To the greengrocer, whose name was Fred.
> He answered, both —— and cool:
> 'That's because they are ——, you fool!'

Solution

lemons, solemn, and *melons.*

Also called Missing Words.

Background Named by Godfrey Bullard and Anthony Creery-Hill after *Hunter* Skinner, who liked such puzzles.

Hypochondriac

👥 Two or more players	✍ Played by speaking
❓ Alphabetical game; cumulative game	✎ No equipment needed

Object To remember an ever-growing alphabetical list.

Procedure This game is similar to I Went to Market, building up a list of words to be remembered by the players. The first player says: 'I went to hospital because I had . . .' and names a disease or illness starting with the letter A, such as *acne* or *anaemia*. The second player has to repeat this and add an illness beginning with B, so this player might say: 'I went to hospital because I had *acne* and *botulism*.' And so on, round the players, with any player dropping out who cannot remember the whole list.

Compare Alphabet Dinner; Grandmother's Trunk; I Packed My Bag; I Went to Market; Traveller's Alphabet.

I'll Haves

㉓ Any number of players

? Punning game

✍ Played by writing or speaking

✎ Played with pencils and paper, or with no equipment

Object To think of suitable food or drink for particular people.

Procedure This is a punning game where the challenge is to think of the sort of nourishment that might be ordered in a restaurant by particular kinds of people. So a barber might like *hare* (hair); a pianist might like *tuna* (as in *piano tuner*); and a golfer might want *tea* (tee).

Example Which food or drink might be appropriate for: (1) a master of ceremonies; (2) a mathematician; (3) an electrician; (4) a boxer; and (5) a jeweller?

Solution (1) *toast*; (2) *pie*; (3) *currants*; (4) *punch*; (5) *carrots*.

106

I Love My Love

⚂ Two or more players
❓ Alphabetical game;
   cumulative game

✏ Played by speaking
✎ No equipment needed

Object To think of appropriate words in an alphabetical series.

Procedure The first player starts the game by saying: 'I love my love with an A, because she is A____,' filling the gap with an adjective beginning with A (*adorable*, for example). The second player says: 'I love my love with a B, because she is B____,' supplying a suitable adjective beginning with B. The game continues round the circle of players. Anyone who cannot think of a suitable adjective has to drop out.

If you wish to make the game more difficult, each player must repeat the preceding adjectives as well as adding a new one, so that the third player might have to say: 'I love my love with a C, because she is *adorable, beautiful,* and *charming.*'

Another variation of the game increases the number of categories for which players have to find adjectives. So the first player might say: 'I love my love with an A, because she is *athletic;* her name is *Angela;* and she comes from *Aberystwyth.*' The number of categories can be increased to include such things as 'I took her to ____' and 'I gave her a ____.' Another possibility is to make each player think of an adverb as well as an adjective: for example, 'I love my love with an A, because she is *amazingly angelic.*'

Compare Minister's Cat.

I'm Going to Take a Trip   *see* Traveller's Alphabet.

Inflated Rhetoric

🎲 Any number of players	✍ Played by writing
❓ Punning game	✏ Played with pencils and paper

Object To write a piece that changes any words which include a number.

Procedure This is a simple punning game in which you produce a piece of comical writing which increases by one the value of any numbers mentioned in it. This includes not only actual numbers but any sound that resembles a number. So you can change 'Once upon a time' to 'Twice upon a time', and 'graduate' to 'gradunine'.

Example Twice upon a time, a young man of the elevender age of fiveteen lived in the twoderful country of Threenisia. He was the son of a powerful potentnine, and he fiverightly hnined his threetor, with a double-minded hninered. So, instead of atelevending his lessons, he second nine his breakfast and then went fifth three seek his fivetune. He met a learned man and said three him: 'I am a Threenisian. Are you two three?' The man replied: 'Yes, but my fivefathers lived in Timbuckthree.' The young man tripled over with laughter, and said: 'How unfivethreennine five them!'

Also called Inflationary Language.

Compare Deflation.

Inflation

🎲 Two or more players	✍ Played by writing
❓ Alphabetical game; cumulative game	✏ Played with pencils and paper

Object To make three-letter words with increasingly higher values.

Procedure This game resembles Centurion, in that the letters of the alphabet are given scores, so that A = 1, B = 2, C = 3, and so on to Z = 26. The first player writes down a three-letter word and, beside it, the score made by its letters. The second player takes one letter from the previous word and writes down another three-letter word with a *higher* score than the preceding word. The players continue adding words which have higher scores than the previous words. The winner is the player who makes the word with the highest total.

Example Chas writes down the word BAG and alongside it the score of ten. Anna writes GET and scores 32. Tony writes down YET, which scores 50. Kate writes down TOY and scores 60. She hopes that nobody can think of a higher-scoring word using one of the letters of TOY, but Chas comes up with WOW which scores 61 and which nobody else can beat, so he wins this round.

Compare Centurion.

Inflationary Language   *see* Inflated Rhetoric.

Initial Answers   *see* Initial Letters (1).

Initial Letters (1)

② ③	Two or more players	✍	Played by speaking
?	Challenge game	✎	No equipment needed

Object To answer questions, using words which start with one's own initials.

Procedure The first player asks the others a question, which they all have to answer in turn, using words which start with the same letters as their own initials. So if the question is 'What is your hobby?', Charles Smith might say 'Collecting Stamps', and Kate Brown might answer 'Keeping Beetles'. The second player then asks a question, which must be answered in the same way, and so on—round the players. Any player who cannot think of a suitable answer has to drop out of that round of the game. If all the players have *three* names, they can use three-word answers.

Also called Initial Answers; Initials.

Compare Hobbies.

Initial Letters (2) *see* Guggenheim.

Initial Proverbs

🔢 Two or more players	✍ Played by writing or speaking
❓ Guessing game	✎ Played with pencils and paper, or with no equipment

Object To guess proverbs from their initials.

Procedure One player gives the other players the first letters of the words of a proverb, and they have to guess what the full proverb is. The first player to guess correctly becomes the next one to set a proverb. Alternatively, one player can read out a list of (say) ten sets of initials which the other players write down and try to solve within a given time.

Example

CHAS: What is this proverb: T–N–S–W–F?

ANNA: *There's no smoke without fire.* What's this: A–R–S–G–N–M?

KATE: *A rolling stone gathers no moss.* Here's another: A–S–I–T–S–N.

TONY: *A stitch in time saves nine.* Now guess this one: C–A–S–S–D.

ANNA: *Coughs and sneezes spread diseases.* Is that really a proverb. . . ?

Initials *see* Initial Letters (1).

Initial Sentences

🔢 Two or more players	✍ Played by speaking
❓ Cumulative game	✎ No equipment needed

Object To build up sentences whose initial letters spell words.

Procedure The players form a circle and the first player starts a sentence by saying a word. Subsequent players in turn add one word each—but the initial letters of the words must build up to form words. So, if the first player says *Tony*, the second player might add *is*, and the third player might add *excruciating*—thus spelling out the word TIE. The game can be played in two ways: players may have to drop out if they cannot continue building up words, or they may drop out if they *complete* a word of four or more letters (compare the game of Ghosts).

Also called Acromania; Acronymia.

Inquisition *see* Questions.

Insertion–Deletion Network

⊘ Any number of players	✍ Played by writing or speaking
? Cumulative game; word-finding game	✎ Played with pencils and paper, or with no equipment

Object To make a sequence of words by alternately adding and deleting letters.

Procedure A word is taken at random (e.g. from a book) or proposed by one of the players. The aim is to add a letter to it to make a new word, and then to remove one letter to make yet another word. Players try to create as long a sequence of words as possible, without repeating any word. If more than one person is playing, they can take it in turns to add or remove letters. Players who cannot think of a suitable word have to drop out of the game until only one person is left. Letters are usually added and removed alternately but players can be allowed to choose whether to add or delete a letter when it is their turn (as in the second of the examples below).

Examples

Fowl–owl–howl–how–show–sow–sown–own–down–don–done–one–once.

Word–world–wold–old–gold–god–goad–gad–ad–wad–wads–was–ways.

In the Manner of the Word *see* Adverbs.

In-Words *see* Words within Words.

I Packed My Bag

⊘ Two or more players	✍ Played by speaking
? Alphabetical game; cumulative game	✎ No equipment needed

Object To remember an ever-growing list of words.

Procedure There are several ways of playing this game:

1. The first player starts the game by saying something like: 'I packed my bag for *Antigua* and in it I put an *aspirin*' (choosing a place-name and an object that both begin with the letter A). The second player then has to think of a sentence with a place-name and object starting with B, for example: 'I packed my bag for *Berlin* and in it I put a *bell*.' And so on, round the players in turn. Anyone who cannot think of a sentence in a reasonable time is out of the game.

2. The first player starts by saying something like: 'I packed my bag and in it I put an *apple*' (choosing an object beginning with A). The second player then has to repeat this sentence, adding an object starting with the next letter of the alphabet, for example: 'I packed my bag and in it I put an *apple* and a *Belisha beacon*.' As the game progresses, each player has to recite an ever-lengthening list of objects. Anyone who cannot remember them all, or cannot think of an appropriate one to add, is out of the game.

3. The game can be played without an alphabetical sequence. In this case, the players simply build up a list of objects starting with any letter at all. So the first player might say: 'I packed my *bag* and in it I put a *toothbrush*.' The second player might say: 'I packed my *bag* and in it I put a *toothbrush* and an *egg-plant*.' Again, any player who cannot remember the whole list has to drop out of the game, and the last person left in is the winner.

In the version of this game called I Went on a Trip, 'I went on a trip to . . .' replaces 'I packed my bag for. . .'

Compare Alphabet Dinner; Grandmother's Trunk; Hypochondriac; I Went to Market; Traveller's Alphabet.

Irish Bulls

②③ Any number of players	✍ Played by writing or speaking
? Challenge game	✎ Played with pencils and paper, or with no equipment

Object To think of absurdly contradictory statements.

Procedure An Irish bull is an obviously ridiculous saying, especially one that contradicts itself—although it may *sound* as if it makes sense. Irish bulls are sometimes called Goldwynisms—after the film producer Samuel Goldwyn, who supposedly said such things as 'A verbal contract isn't worth the paper it's written on.' The aim of the game is to think of Irish bulls, either for one's own amusement or to share with other players.

Examples

His mother had no children.

If you don't receive this letter, please let me know.

I'm glad I hate cabbage because, if I liked it, I'd hate it, and I can't stand cabbage.

I sold my treasure-chest, so that I'd have something to put in it.

I took so much medicine that I was sick a long time after I got well.

I looked, and there he was—gone!

Compare Malapropisms; Oxymorons.

Isograms

23 Any number of players
? Challenge game; letters game

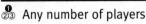 Played by writing
Played with pencils and paper

Object To find words consisting of letters used the same number of times.

Procedure Isograms are words that consist of letters used a particular number of times—usually the same letters used either once, twice, or three times. It is easy to find short words consisting of letters used only once, like *dog, cat,* or *turnip,* but the challenge is to find a long word in which no letter is repeated. Players can try to write down longer and longer words of this kind, perhaps moving through *comedian* to *discountable* and *dermatoglyphics.*

Players can also look for words which consist of the same letters repeated twice or three times. Two-letter isograms include *gee-gee, tartar, reappear,* and *intestines. Deeded* is a three-letter isogram.

Isosceles Words

23 Any number of players
? Anagrams game; cumulative game

 Played by writing
Played with pencils and paper

Object To build a triangle out of anagrams that increase or decrease in length.

Procedure This game is a form of Transadditions and Transdeletions. Players write a letter at the top of a page, add another letter to it, and write a two-lettered word below it, add another letter and (if necessary) rearrange the three letters to make another word which is written beneath the other two. In this way, a pyramid is made out of words of ever-increasing length. Alternatively, players can start with a long word and remove one letter at a time, rearranging the letters each time to make a pyramid of words that gradually decrease in length. If a shape is made that first increases and then decreases in size, the game is called One to One.

Examples

A	TRIFLES
AT	STRIFE
HAT	FIRST
HEAT	RIFT
THANE	FIR
ANTHER	IF
UNEARTH	I
URETHANE	

Also called Pyramids.

Compare Shrink Words; Transadditions; Transdeletions.

I Spy

Two or more players   Played by speaking
Guessing game   No equipment needed

Object To guess a word that starts with a particular letter.

Procedure One of the players says: 'I spy, with my little eye, something beginning with . . .' and gives a letter of the alphabet that starts a word for something that everyone present can see. The other players then try to guess what the object is. Whoever guesses correctly becomes the next player to choose an object. Alternative versions of the game give either a word that the object rhymes with, or the colour of the object.

Example

KATE: I spy, with my little eye, something beginning with C.

CHAS: Is it *curtain*?

KATE: No.

ANNA: Is it *chest-of-drawers*?

KATE: No.

TONY: Is it a *champagne bottle*?

KATE: No, I can't see one.

CHAS: Is it *chin*?

KATE: No.

ANNA: Is it the *cat*?

KATE: Yes!

I Went on a Trip *see* I Packed My Bag.

I Went to Market

Two or more players

? Alphabetical game; cumulative game

Played by speaking

No equipment needed

Object To remember an ever-growing alphabetical list.

Procedure The first player says something like: 'I went to market and bought an *aubergine*' (choosing an object that starts with the letter A). The second player repeats this sentence, adding something beginning with B, for example: 'I went to market and bought an *aubergine* and a *billy-goat*.' And so on round the players, with each player adding a new word. Anyone who cannot remember the items previously listed, or who cannot think of a suitable word starting with the next letter, is out of the game.

In the version of this game called I Went to the Store, the phrase 'I went to the store' replaces 'I went to market'.

Compare Alphabet Dinner; Grandmother's Trunk; Hypochondriac; I Packed My Bag; Traveller's Alphabet.

I Went to the Store *see* I Went to Market.

Izzat So? *see* Perverbs.

116

J

Jarnac

 Two players (or two teams)
? Word-building game

🖎 Played on a board
🖎 Played with a boxed set of two boards and a bag of 144 letter tiles

Object To form words on a board.

Procedure Each player (or each team) takes a board (which has a grid of nine-by-eight squares drawn on it). The first player draws six tiles from the bag of letters, and tries to place a word of at least three letters horizontally on the first line of the grid. If the first player succeeds in doing this, the same player draws one more tile from the bag, and tries either to change the word on the grid by adding one or more letters, or to place another word on the second line of the grid. This player can continue drawing one letter at a time and trying to make or change a word, until no more words can be made. If the second player sees that the first player could have made another word, the second player shouts 'Jarnac' and takes those letters from the first player's board and puts the word on the first line of his own board. Play then passes to the second player, who goes through the same procedure as the first player. After the first round, each player takes turns to draw a new letter and try to make a new word. Alternatively, players can choose to exchange up to three of their tiles for three others from the bag. The game ends when one player has made a word of any length on all eight lines across his or her board. Scores are then calculated on the basis of the numbers at the top of each 'down' line of the grid: three-letter words score nine; four-letter words score 16; five-letter words score 25; six-letter words score 36; seven-letter words score 49; eight-letter words score 64; and nine-letter words score 81. The winner is the player with the highest score, whether or not that player has a word on every line.

Background A boxed game, copyright J. W. Spear & Sons plc.

Jaxsquare

👥 Any number of players
❓ Grid game
✍ Played by writing
✎ Played with pencils and paper

Object To make a word square from the letters in a phrase.

Procedure Jaxsquares are puzzles in which players are given a 25-letter phrase and asked to make a word square using those letters. To create a Jaxsquare, it is possible to choose a 25-letter quotation or phrase and make the word square out of its letters, but it is usually easier to make a word square first and then form the 25-letter phrase.

Example Make a word square from the letters in the phrase: SERVE TABLE ALL SUMMER IN CLUB. The answer is:

C	L	I	M	B
R	E	N	A	L
U	V	U	L	A
M	E	R	E	S
B	L	E	S	T

Compare Crossword Game; Word Squares (1).

Background Invented by John Matthews.

Jotto

👥 Two players
❓ Guessing game
✍ Played by writing
✎ Played with pencils and paper

Object To guess a word, usually a five-letter word, chosen by one's opponent.

Procedure This game is played rather similarly to Crash, except that the guesser is told when *any* of the letters in the 'test' word is the same as *any* of the letters in the hidden word. So, if the mystery word is *crash* and the guesser suggests *break,* the guesser is told that two letters in *break* correspond to letters in *crash.*

Also called Giotto; Secret Word; (as a boxed game) Word Master Mind or Words Worth (both copyright Invicta Plastics Ltd).

Compare Convergence; Crash; Double Jeopardy.

Journalism

② ③ Two or more players	✍ Played by writing
？ Cumulative game	✎ Played with pencils and paper

Object To answer a question, using a particular word in the answer.

Procedure Each player is given three slips of paper. On the first slip, each player writes a question. On the second slip, each player writes a single word. These slips are collected into two separate piles and jumbled up. Each player is then given one slip from each of the two piles, and has to answer the question on one slip by writing down a sentence that includes the word on the second slip. The answers are then read out. This game often demands ingenuity from the players, who may have to answer a question like 'What is the square root of 64?' with a sentence that includes the word *petunias.*

Jumble

② ③ Any number of players	✍ Played by writing
？ Anagrams game	✎ Played with a previously prepared puzzle, plus pencils and paper

Object To solve four anagrams, and then a fifth arising from the four.

Procedure A Jumble consists of several words whose letters have been jumbled. One or more of the letters in each word is circled. The player has to solve all the anagrams, and then discover an extra anagram in the circled letters.

Example The five scrambled words are: F(E)U L Y E, C I P O L(Y), S U B(R) I F, G L E B A(M), and T O (II) O T. These can be rearranged to make the words *eyeful, policy, bruise, gamble,* and *tooth.* The circled letters have the clue 'The usual way to finish a poem' and they can be turned into the word *rhyme.*

Jumbled Proverbs

👥 Three or more players	✏️ Played by writing or speaking
❓ Guessing game	✎ Played with pencils and paper, or with no equipment

Object To unscramble jumbled proverbs.

Procedure One of the players prepares a list of proverbs whose words have been jumbled. This player then either reads them aloud to the other players or hands them out on sheets of paper. The other players have to put each proverb into its correct form. For example, the players might be given 'Empty noise make the most vessels' and they have to unscramble it into: 'Empty vessels make the most noise.' To make the game more difficult, the first player can hand out lists of proverbs in which the *letters* of the words are jumbled, so that the other players would have to turn 'Loofs shur ni herew angles fare ot trade' into 'Fools rush in where angels fear to tread.'

Jumbled Words *see* Anagrams (1).

Junctures *see* Anguish Languish.

Just a Minute

👥 Three or more players
❓ Challenge game

✍ Played by speaking
🕙 Played with a clock or other timer

Object To continue speaking for one minute on a chosen subject.

Procedure One player is chosen as the umpire, and sets the other players subjects which they have to speak about for one minute—without hesitation, repetition, or deviation from the subject. A speaker who breaks any of these rules can be challenged by another player and, if the umpire upholds the challenge, that player continues on the same subject for the rest of the minute. The winner of each round is the player who is speaking as the minute expires. Alternatively, points can be scored for successful challenges and for being the speaker who completes a minute's talk.

Example Kate is the umpire, and she asks Anna to talk for one minute about *Dustbins*.

ANNA: Dustbins are large metal containers for rubbish which are collected by dustmen, who empty the rubbish . . .

CHAS: Challenge! Repetition of the word *rubbish*.

KATE: Challenge accepted. Chas will now continue speaking about *Dustbins* for 52 seconds.

CHAS: Dustbins are not always made of metal. Nowadays many of them are plastic and . . . er . . .

TONY: Challenge! Hesitation.

KATE: Accepted. Tony will continue speaking for 43 seconds about *Dustbins*.

TONY: Our dustbin is at the front of the house, and the cats often get inside the bin and drag out things to eat, which usually makes them sick all over the kitchen floor . . .

ANNA: Challenge! Deviation: he is talking about cats and kitchens, not dustbins.

KATE: Accepted. Anna will continue speaking about *Dustbins* for another 28 seconds. (*And so on.*)

Also called One Minute, Please.

Compare Ad Lib.

Background Broadcast for many years on BBC Radio 4 as a popular game played by four people and an umpire.

K

Kangaroo Words

👥 Any number of players	✍ Played by writing or speaking
❓ Word-finding game	✎ Played with pencils and paper, or with no equipment

Object To find words hidden inside longer synonymous words.

Procedure A list is prepared beforehand of kangaroo words: that is, words that contain embedded inside them other words which mean the same thing, such as *devilish* (which has *evil* inside it) or *instructor* (which has *tutor* inside it). Other examples are asSUREd, obSErvE, bRIM, ROtUND, FRAgILe, exISts, transgresSIoN, FabrICaTION, masticATEd, and faLsitIES. Players are then given the words one at a time, and they have either to say or write down the synonyms they find therein. The winner is the person who spots the largest number of words.

Also called Marsupials.

Kan-U-Go

👥 Two to seven players	✍ Played with cards
❓ Word-building game	✎ Played with a boxed set of 58 lettered cards and two 'Kan-U-Go' cards

Object To form words with lettered cards.

Procedure Players are dealt a number of cards, depending on how many people are playing (for example, if two are playing, they receive 12 cards each; if seven are playing, they receive seven cards each). The remaining cards are placed face-downwards on the table and the top card is turned up and placed beside the pack, to start a discard pile. The player to the left of the dealer tries to place on the table a word of two, three, or four letters. Alternatively this player can take the top card from the pack on the table or the discard pile, placing one of his or her own cards on the discard pile. Play continues with each player trying to make new words by adding letters to the words on the table, forming an interlocking grid pattern as in Scrabble. The two 'Kan-U-Go' cards can represent any letter but can only be laid down (or thrown away) when players are playing their *last* card or cards. The winner of a round is the first player to lay down all his or her cards. The other players count their scores from the cards remaining in their hands. The game ends when any player reaches a score of 100, and the winner is the player with the lowest score.

Background A boxed game, copyright Waddingtons Games Ltd, under licence from Jarvis Porter Ltd. Kan-U-Go is a registered trade mark of Waddingtons Games Ltd.

Keats and Chapman *see* My Word (1).

Keyword (1)

②③ Two or more players	🖾 Played by writing
❓ Word-finding game	🖎 Played with pencils and paper

Object To discover the word from whose letters other words are made.

Procedure This game is the reverse of Words within Words. One player prepares a list of words made from the letters of a word. The other player(s) then have to work out what the secret word is.

Example Tony gives Chas the following words: *crate, brace, tea, late, real, rat,* and *create,* and he tells Chas that they were all made from the letters of a nine-letter word. Chas eventually works out that the word is *celebrate.*

Keyword (2) *see* Password; Words within Words.

Key Words *see* Words within Words.

Knickers *see* Sausages.

Knight's Tour

 Any number of players   ✍ Played by writing
? Grid game; word-finding   ✎ Played with a previously
   game   prepared grid

Object To find a hidden word or words in a grid.

Procedure This game is very much like Word Search. However, instead of moving around the grid horizontally, vertically or diagonally, players move in the same way as the knight in chess: one square forward and then two squares sideways, or two squares forward and one square sideways. Words are hidden in a grid which is prepared by another player.

Example Work your way around this grid in 'knight's moves' to spell out the title of a play by George Bernard Shaw:

U	R	A	
N	N	S	E
M	P	A	N
M	D	A	

Solution

Man and Superman.

125

Knock-Knock

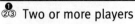

⊘ Two or more players	✍ Played by speaking
? Guessing game; punning game	✎ No equipment needed

Object To devise or answer jokes that use punning names.

Procedure Knock-knock jokes depend on names which include (or lead to) a pun. One player says 'Knock-knock', as if knocking at a door; another player asks: 'Who's there?' and the first player replies with a name. The second player either guesses the pun or asks for the answer.

Examples

CHAS: Knock-knock.
ANNA: Who's there?
CHAS: Lemmy.
ANNA: Lemmy who?
CHAS: Lemmy in and I'll tell you.

KATE: Knock-knock!
TONY: Who's there?
KATE: Shelby.
TONY: Shelby who?
KATE: Shelby coming round the mountain when she comes . . .

Knowledge List *see* Guggenheim.

Kolodny's Game

👥 Two or more players	✍ Played by speaking
❓ Guessing game	✎ No equipment needed

Object To guess the rules which govern the answers that are given to questions.

Procedure One player is chosen as the question-master, and thinks of a secret rule that will be followed in answering questions from the other players. For example, the rule might be: 'All questions that start with a two-letter word will be answered *yes*; all other questions will be answered *no*.' The other players then ask questions in turn, and the question-master replies only *yes* or *no*, observing the secret rule. From the answers given, the other players try to guess what that rule is.

Example Tony thinks of the rule that questions of fewer than five words will be answered *yes*; questions of five words or more will be answered *no*.

KATE: Are you as beautiful as you look?
TONY: No.
ANNA: Are you silly?
TONY: Yes.
CHAS: What is the time?
TONY: Yes. (*And so on, until the other players guess the rule that Tony is using, or collapse from exhaustion.*)

Background Game invented by David Greene Kolodny.

Kriss Kross

👥 Any number of players	✍ Played by writing
❓ Grid game	✎ Played with a previously prepared grid

Object To solve a puzzle of interlocking words.

Procedure A Kriss Kross is rather like a crossword puzzle, except that the diagram is not a regular square but a series of interlocking spaces in which letters have to be entered. A list is supplied of the words to be filled in, and players simply have to put them into the correct squares. Usually the words are from a particular category, like flowers, birds, authors' surnames, etc. People who create these puzzles can make them more difficult by interserting each word with several other words, and using words which contain the same number of letters. In the example below, the words all contain six or seven letters, and the category is 'animals'.

Example Fit these animals into the grid: *aurochs, cougar, coyote, dragon, fennec, gerbil, gopher, grampus, heifer, kitten, lambkin, panther, plover, polecat, possum, rhesus, seagull, serpent, sponge, teledu, tigress, wapiti,* and *weasel.*

Solution

L

Laddergrams *see* Doublets.

Laddergraphs *see* Doublets.

Last and First *see* Geography.

Last Shall Be First *see* Geography.

Last Word (1)

❶ Two or more players	✍ Played by writing
❓ Grid game	✎ Played with pencils and paper

Object To build up words in a grid.

Procedure A sentence is chosen at random from a book, and its first nine letters are written in the nine central squares of a nine-by-nine (or larger) grid. Players alternately add single letters which make two or more words, horizontally, vertically, or diagonally. The scores are calculated by multiplying together the number of letters in all the words made with one move. The game ends when there is a letter in at least one square along each side of the grid.

Example Anna took a *Complete Shakespeare* off the shelf and opened it at random, copying the first nine letters of the chosen sentence into the centre of the grid, thus:

T O B
E O R
N O T

Chas places an E after *not* to make *note* and (diagonally) *ore*, scoring 4 × 3 points = 12. Kate adds a T under *ten* to make *tent* and (diagonally) *rot*, scoring 12. And so it continues, with longer words becoming possible, and high scorers trying to reach the outer edge of the grid to finish the game before lower scorers catch them up.

130

Last Word (2)

👥 Three or more players
❓ Anagrams game; word-building game
✏️ Played by speaking
🖊️ Played with a set of letters

Object To make words from letters presented one at a time.

Procedure One player becomes the question-master, and takes charge of a set of letters taken from a set of Scrabble or Lexicon, or written beforehand on cards. The question-master presents the other players with a sequence of three letters, chosen at random. The question-master then shows (and calls out) a fourth letter. The players try to use the four letters to make a word (which they shout out), either by swapping the fourth letter for one of the other letters—and making a three-letter word—or by rearranging all four letters. So, if the first three letters displayed were P–C–E and the fourth letter was A, a player might swap the A for the P and make *ace* or use all four letters and make *pace*. The question-master keeps the score, which gives players one point for every letter in a word they make. When seven letters have been displayed, players can shout out extra letters to make increasingly long words, until nobody can think of any more.

Example The question-master turns up the letters A, L, and E to start with. Then the question-master turns up (and calls out) the letter T. Kate shouts out the word *tale* and scores four points. The question-master turns up the letter P, and Chas shouts out *pleat* to score five points. The question-master turns up A, and Tony shouts out *palate* to score six points. The question-master turns up C, and Anna shouts out *placate*, scoring seven points. The question-master then invites the players to add an eighth letter to make a new word, and Kate shouts out N, making *placenta* . Nobody can think of another letter to make a nine-letter word, so the question-master starts a new round.

Last Word (3) *see* Trackword.

Leadergram *see* Double-crostic.

Leading Lights *see* Hobbies.

Letter Auction

 Two or more players

? Word-building game

 Played on a table

Played with counters or matches, and a set of letters

Object To make words from letters that are auctioned.

Procedure One player becomes the 'auctioneer' and takes the set of letters, which can be taken from a set of Scrabble or Lexicon, or made by drawing letters on pieces of cardboard. The set should contain between about 50 and 100 letters, with more of the common letters like E and S than the uncommon letters like K and Z. Each player is given the same number of counters or matches to use as 'money' in the game. The auctioneer turns up the first letter and offers it for sale, as in an auction. The player who bids the most for that letter gets it, and pays the auctioneer the price that has been bid. As the auction continues, the players try to use their letters to form words on the table in front of them. Players can also offer to buy letters from other players if they need particular letters to complete a word.

When all the letters have been auctioned off, the players have an agreed time-limit to complete their words, and then the scores are worked out. If letters are used from Scrabble or Lexicon, the scores can be totalled from the scores printed on each tile or card. If a home-made set of letters is used, scores can be awarded on the basis of one point for every letter used in an acceptable word. The scores for the letters are added to the number of counters or matches that each player has not 'spent', and the winner is the player with the highest score.

Letter Bank

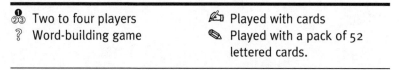

🔵 Any number of players	✍ Played by writing
❓ Word-finding game	✏ Played with pencils and paper

Object To find a word or phrase that is made from the letters of a shorter word.

Procedure A letter bank is a puzzle, sometimes in prose but often in verse, giving clues to a 'keyword' and another word or phrase. The keyword has no letter repeated; the other word or words are spelt with only the letters of the keyword—but those letters may occur more than once. So the keyword might be *imps* and the second word could be *Mississippi*, or the keyword might be *bathers* and the phrase from its letters could be *The Three Bears*. The keyword is often given as SHORT or ONE; the resulting word or phrase as LONG or TWO.

Example

> He ruled in Argentina's court,
> His wife was more famous in song:
> Well-dressed was the woman called SHORT,
> And she certainly knew how to LONG.

If the players need clues, they can be told that the SHORT word has five letters and the LONG word has eight letters.

Solution *Evita* and *titivate*.

Letter by Letter   *see* Doublets.

Letter Change   *see* Doublets.

Lexicon

🔵 Two to four players	✍ Played with cards
❓ Word-building game	✏ Played with a pack of 52 lettered cards.

Object To make words with lettered cards and to get rid of your cards as quickly as possible.

Procedure Ten cards are dealt to each player. The rest of the cards are placed in the centre of the table, face-down. The top card is placed face-up next to the pile. Each player in turn does one of four things: (*a*) makes a word from the letters he or she holds, and places it face-up on the table; (*b*) takes one card from one of the piles on the table and discards another card, placing it on the face-up pile; (*c*) adds a card or cards to a word already on the table, or (*d*) exchanges one or more of his or her cards for one or more of the cards in words already on the table, making a new word or words. When one player gets rid of all his or her cards, the other players total the numbers on the cards they are still holding. The first player who reaches 100 or more, drops out of the game—and so on, until only one player is left as the winner.

The cards in the pack consist of the following:

Letter	Number of cards in pack	Score for each card
A	4	10
B	1	2
C	1	8
D	1	6
E	4	10
F	1	2
G	1	4
H	3	8
I	4	10
J	1	6
K	1	8
L	3	8
M	1	8
N	1	8
O	3	8
P	1	8
Q	1	4
R	3	8
S	3	8
T	3	8
U	3	8
V	1	6
W	3	8
X	1	2
Y	1	4
Z	1	2

There is also one 'master card' (used like a 'joker'), with a score of 15.

Lexicon can also be played (as Lexicon Criss-Cross) as a grid game, with players writing words in a five-by-five grid from the letters revealed by turning up the cards in succession, until 25 letters have been revealed. Players score ten points for making a five-letter word, seven points for four-letter words, three points for three-letter words, and one point for two-letter words.

Another variation is Lexicon Riddance, in which the players are dealt seven cards each, and have to make words by adding letters to cards turned face-up in succession from the pile of spare cards.

A further variation is Lexicon Clock Patience, for one player. Twelve cards are placed face-up in a circle, with another card face-up in the centre of the circle. The player tries to make twelve four-letter words round the circle (reading inwards towards the centre of the circle), and another four-letter word in the centre. The player turns up every third card in the pack and tries to place it in such a position as to help make a word.

Background A boxed game, copyright Waddingtons Games Ltd. It is a 'classic playing card word-game which has been a firm favourite for more than three generations'. Lexicon is a registered trade mark of Waddingtons Games Ltd.

Licence Plate Game *see* Catchword.

Limericks *see* Group Limericks.

Limericksaw *see* Group Limericks.

Lipograms

②③ Any number of players	✍ Played by writing
⸮ Challenge game	✎ Played with pencils and paper

Object To write something without using a particular letter.

Procedure Lipograms are pieces of writing which deliberately omit one letter of the alphabet. It is easy enough to write something without using such letters as Q or Z but the challenge is to write a sensible piece without using a more common letter. Start, for example, by writing something without using the letter T; then the letter S; then try E. If this is not difficult enough, try writing poetry rather than prose.

Compare Univocalics.

Logograms *see* Logogriphs.

Logogriphs

👥 Any number of players	✍ Played by writing
❓ Anagrams game; letters game	✏ Played with pencils and paper

Object To guess a word from anagrams for which clues are given.

Procedure A logogriph is a puzzle in which a mystery word goes through several transformations, some or all of its letters being rearranged to make several new words. Clues are given to each of the words. Logogriphs can be in the form of verses or in prose.

Example 'I am a seven-letter word that you may find down at the theatre. My 1,2,3,4 is short; my 3,2,7 may be long; and my 4,3,5,6,7 is a set of carriages.' (The answer is *curtain*.)

Also called Logograms; Numericals.

Lost Consonants *see* Misprints.

Lucky Ladders

👥 Two players (or two teams)	✍ Played by writing or speaking
❓ Guessing game	✏ Played with pencils and paper, or a blackboard, etc.

Object To trace a series of words associated in sense.

Procedure Players are given by the question-master a prearranged pair of words: one word begins a chain of word associations; the other ends the chain. The players have to work their way down the chain, being helped by being given letters one at a time. The winner is the first player to complete the chain.

Example The players are given this 'ladder' to solve:

B A T H

_____

_____

_____

_____

_____

D E G R E E

Chas starts by trying to guess the word beneath *bath*, which must be associated in some way with *bath*. He asks to be given a letter, and he is told the first letter is J. Chas cannot guess it, so it is Kate's turn. She asks for another letter and is told A. She guesses correctly that the second word is *jacuzzi*. She then asks for the first letter of the third word, and is told it is J. She cannot guess this word, so the turn passes back to Chas, who asks for another letter and is given E. From this (and from the connection with *jacuzzi*), he guesses that the third word is *jet*. And so on, until the remaining words in the ladder are filled in: *black, blue,* and *university*—leading to *degree*.

Background Popularized as a quiz show on ITV introduced by Lenny Bennett.

Lynx

👥 Two or more players   ✍ Played by writing
❓ Grid game   🖊 Played with pencils and paper, and a crossword grid

Object To score points by entering words in an empty crossword grid.

Procedure An unanswered crossword is taken from a newspaper or magazine. Each player in turn has to write a word into a suitable place in the grid. As other words are added, they must fit in with the words that are already there. Every word must link with at least one of the words that is already in the grid. Each word scores points for the number of letters it contains, multiplied by the number of other words it is linked with (alternatively, it can score the number of letters *added to* the number of words it links with). Play continues until nobody can think of any more words to enter. The winner is the player with the most points.

Example Anna writes the word *examples* in the grid, and scores seven points. Kate adds *slight*, using the L in *examples* and scoring six points. Anna uses the G in *slight* to start *gluttonous*, scoring ten points. Kate links the S of *examples* and the U of *gluttonous* to make the word *shutter*, scoring 14 points (seven for the letters in *shutter*, multiplied by two for the number of words it links). And so on.

Compare Alphacross.

Background Game invented by David Parlett.

M

Making Words *see* Words within Words.

Malapropisms

②③ Any number of players	✍ Played by writing or speaking
？ Punning game; wordplay game	✎ Played with pencils and paper, or with no equipment

Object To use words that are mistaken for correct words.

Procedure Malapropisms are words or phrases which are used in the wrong context because they are mistaken for similar words. They are named after Mrs Malaprop in Sheridan's play *The Rivals*—a character who would say such things as 'An allegory on the banks of the Nile' (meaning 'an alligator') and 'Illiterate him, I say, quite from your memory' (meaning 'obliterate'). Mistakes like these are often called howlers, especially if they are made by schoolchildren or students.

Used as a game, malapropisms can be spoken by one player to another, who has to guess the correct word or phrase. Or you can simply make up malapropisms for your own and other people's enjoyment.

Example

ANNA: Can you correct the malapropism in this sentence? 'Comparisons are odorous.'

KATE: It should be 'odious'. Can you spot the mistake in this sentence? 'You must keep your nose to the tombstone.'

CHAS: It should be 'grindstone'. There are two errors here: 'He had an operation on his prostrate gland and then his leg went sceptic.'

TONY: It should be 'prostate gland' and 'septic'. Talking of medical problems, I've got very close veins in my leg.

ANNA: You mean 'varicose veins'.

Compare Irish Bulls; Misprints.

Marsupials *see* Kangaroo Words.

Mascots *see* Bullets.

Metagram

Any number of players	Played by writing
Guessing game; word finding game	Played with a previously prepared puzzle

Object To guess a word from other words which are almost the same.

Procedure A metagram is a challenge to discover a word from clues to other words which have one letter different from the secret word.

Example This example comes from F. Planche's *Guess Me* (published in 1872): I am a word of four letters. Change my first, and I am to wish for; my second, and I am a road; my third, and I am a tree; my fourth, and I am a fish.

Solution *line* (*pine–lane–lime–ling*).

Minister's Cat

Two or more players	Played by speaking
Alphabetical game	No equipment needed

Object To think of suitable adjectives in an alphabetical sequence.

Procedure This game is rather like I Love My Love. The first player says something like 'The minister's cat is an amazing cat,' choosing an adjective starting with A to describe the cat. The next player has to think of a different adjective starting with A to describe the cat (e.g. 'The minister's cat is an artful cat')—and so on round the players. For the second round, the adjectives have to begin with B, and so on through the alphabet. Anyone failing to think of a suitable adjective has to drop out of the game. To make the game more difficult, players can also be made to name the cat ('The minister's cat is an arrogant cat and its name is Agamemnon') or they may have to list all the adjectives so far used for a particular letter, as in the example below.

Example

ANNA: The minister's cat is an adorable cat.

KATE: The minister's cat is an adorable, artistic cat.

TONY: The minister's cat is an adorable, artistic, awful cat.

CHAS: The minister's cat is an adorable ... er ... artistic, awful, antagonistic cat.

KATE (*starting the second round*): The minister's cat is a beautiful cat.

TONY: The minister's cat is a beautiful, baggy cat.

CHAS: The minister's cat is a beautiful, baggy, bandy cat.

ANNA: The minister's cat is a beautiful, baggy, bandy, besotted cat.

Also called Parson's Cat; Preacher's Cat; Vicar's Cat.

Compare: I Love my Love.

Mischmasch

🔴 Two or more players

❓ Word-finding game

✍ Played by writing or speaking

✎ Played with pencils and paper, or with no equipment

Object To think of a word that contains a particular set of letters.

Procedure One player gives the other(s) a set of two or more letters. The other player(s) have to think of a word that contains those letters together. Whoever thinks of a suitable word becomes the next person to propose a 'nucleus'.

Example Tony proposes the letters *mew* to the other players. Nobody can think of a suitable word until Kate comes up with *mewling*. Other possibilities include *mews, smew* (a kind of duck), and—more cleverly —*homeward, homework,* and *somewhere.*

Compare Superghosts.

Background Game invented by Lewis Carroll in 1880.

Misprints

🔴 Any number of players

❓ Punning game

✍ Played by writing

✎ Played with pencils and paper

Object To explore the possibilities offered by misprints.

Procedure Misprints can be a source of amusement, as when a newspaper described a policeman as 'a member of the defective farce' instead of 'the detective force'. For a game based on misprints, players can try changing one letter—or more—in a word or phrase (for example, a book title, a proverb, a person's name) with incongruous or punning results. Such changes may require explanations or definitions.

Examples

Happily even after.

God help those who help themselves.

I'm in the wood for love.

Workers arise! You have nothing to lose but your chairs!

Oliver Twit: Charles Dickens' famous novel about a foolish boy.

The Lone Rager: the last angry man.

The Wild Brunch: a violent western about cowboys arguing over a late breakfast.

Also called Lost Consonants; One-Letter Omissions.

Compare Malapropisms.

Missing Letters

② Two or more players	✍ Played by writing
❓ Guessing game	✎ Played with pencils and paper

Object To guess incomplete words with the help of clues.

Procedure This is rather like a crossword puzzle without the grid. One or more of the players prepare a list or lists of words with some of their letters omitted. Each word has a clue, and the other players must use these clues to try and guess the correct words.

Example Chas hands out to the other players this list of eight words with some letters missing and clues attached:

(1) L**E (you can't live without it).

(2) *CA*L*T (ribbons for your hair—or Will).

(3) T**NT* (a score or five quartets).

(4) *S*E*T (rise, like money).

(5) *O*N*C (this medicine weighs a lot plus 99).

(6) C**F*E (you can have it in black and white).

(7) L*S** (a rope around a girl with nothing).

(8) U* (old city sounds uncertain).

Solution (1) *life*; (2) *scarlet*; (3) *twenty*; (4) *ascent* (as + cent); (5) *tonic* (ton + the Roman numeral IC = 99); (6) *coffee*; (7) *lasso* (lass + o); (8) *Ur* (sounds like *er*).

Missing Middles *see* Tops and Tails.

Missing Vowels

⓸ Two or more players	✍ Played by speaking
❓ Guessing game	✎ No equipment needed

Object To guess words with their vowels omitted.

Procedure One player thinks of a word and spells it out to the other player or players, omitting all its vowels. If desired, the questioner can say what kind of word it is: an animal, a flower, etc. The other player or players try to spell out the full word. If they have difficulty, they can be given clues. They score a point if they give a proper word, even if it is not the word that the questioner originally had in mind.

Another way of playing the game is for one player to ask another to spell a word without naming any of the vowels. So one person might ask another to spell the word *elephant* and the answer would be L–P–H–N–T.

Example

ANNA: Can you add some vowels to these consonants to make an animal: B–B–N?

KATE: *Baboon.* Can you make a bird by adding some vowels to these consonants: W–L?

TONY: *Wolf?*

CHAS: That's not a bird! I think the answer is *owl*. Can you add some vowels to the consonant K to make a bird?

ANNA: I know! *Auk.*

Also called Vowel Play.

Missing Words *see* Huntergrams.

Monkey *see* Ghosts.

Monosyllabic Verse

Any number of players

Challenge game; poetic game

Played by writing

Played with pencils and paper

Object To write verse in monosyllables.

Procedure The aim of this game is to write a poem in words of one syllable. This may seem difficult but it was achieved by such famous poets as George Herbert:

> Come, my joy, my love, my heart:
> Such a love, as none can move;
> Such a love, as none can part;
> Such a heart, as joys in love.

You can try this alone or in a group (perhaps with each person writing one line of the poem). Another challenge is to write a piece of prose using only words of one syllable.

Monosyllables

Two or more players

Challenge game

Played by speaking

No equipment needed

Object To speak in words of only one syllable.

Procedure The players form a circle and each in turn asks the next player a question. The question and the answer must only use words of one syllable, and the answer cannot be *yes* or *no*. Any player is out who uses a word containing more than one syllable, and the winner is the last remaining player.

Example

ANNA (*to Tony*) How do you like rain?

TONY: It is not nice. (*to Kate*) Are you old?

KATE: Not at all. (*to Chas*) Why are you so tall?

CHAS: I was born that way. (*to Anna*) Where is France?

ANNA: It is in the world. (*to Tony*) How much cash do you earn?

TONY: Mind your own business. Oh dear! (*He is out.*)

Also called One Syllable.

Compare Short Story.

Mosaics *see* Centos.

Mosaic Verses *see* Centos.

Multiwords *see* Words within Words.

My Name is Mary

Three or more players	Played by speaking
Active game; cumulative game	Played with a doll or other object

Object To build up a confusing series of names.

Procedure The players form a circle, and an object like a doll or toy animal is passed round the circle. The first player hands the object to the second player and states his or her own name as well as the name of the object, for example: 'My name is Mary and this is Fred.' The second player hands the object to the third player and says: 'My name is Tony and Mary says that this is Fred.' The third player hands it on to the fourth player, saying: 'My name is Kate, and Tony says that Mary says that this is Fred.' And so on, with ever-increasing complexity. Anyone who cannot remember the whole list of names has to drop out of the game.

Compare A What?

My Word (1)

Two or more players	Played by speaking
Punning game	No equipment needed

Object To devise ridiculous puns to explain the origin of phrases.

146

Procedure One player suggests a well-known phrase, idiom, or quotation and another player explains its origin, using terrible complicated puns and sometimes perverting the original words of the phrase.

Example

KATE: Can you tell me the origin of the catch-phrase 'What's up, Doc?'

CHAS: It derives from the time when James Watt, the inventor of the steam engine, was ill. The doctor told James's servant to inform him when Watt felt better. The next day, the servant sent the doctor a message, reading: 'Watt's up, Doc!'

ANNA: What is the origin of the proverb 'Don't put all your eggs in one basket'?

TONY: It comes from the time when a French hotel manager hired a band from the Basque region of northern Spain to play for a dance at his hotel. The musicians all arrived by coach but the coach was late, so they all hurried to get into the hotel as quickly as possible. In their haste, they unfortunately got stuck in the revolving door and were unable to get out in time to play for the dance. The moral is: 'Don't put all your Basques in one exit!'

Also called Keats and Chapman; Orrible Origins; Puns in Perpetuity.

Background Popularized by Frank Muir and Denis Norden in the BBC Radio programme *My Word.*

My Word (2)

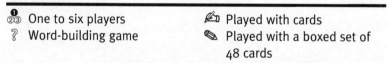

⊕
②③ One to six players

? Word-building game

✍ Played with cards

✎ Played with a boxed set of 48 cards

Object To make four-letter words using cards on which are printed pairs of letters.

Procedure The dealer deals three cards to each player (or four cards if there are only two or three players). The remainder of the pack is placed face-down in the centre of the table and the top card is turned face-up. Every card has four pairs of letters printed on its face: one pair each at the top, bottom, and on each side of the card. Each player, when it is his or her turn, can either: (1) put one of his or her cards at the bottom of the pack in the middle of the table and take the top card; (2) place a card alongside a card already on the table so as to form a word (cards must be placed with their long edges or their short edges side-by-side); or (3) put a card on *top* of a card already on the table, so long as it makes new words with every card it touches. Each pair of letters on the cards has a score printed alongside it, so that players can work out their scores as they make new words. Some cards have symbols instead of letters: these can represent any letter the player chooses. Using these symbols triples the score for the word made with them. There should not be more than six cards side-by-side in any direction, but players can agree to drop this rule. The winner is the player with the highest score when the pack is finished and either one player has played all his cards or no player can make any more words. The game can also be played by one player alone, using four cards and taking new ones from the central pack.

Background Copyright Waddingtons Games Ltd.

N

Name Game

23 Three or more players   🖎 Played by writing and speaking

? Guessing game   🖎 Played with pencils and paper, plus a container

Object To guess the names of famous people.

Procedure The players form a circle round a table. Each of the players writes down on slips of paper the names of a number (say four) of famous people: either real or fictitious. The slips of paper are folded, put into a basket or other container, and shaken well to mix them up. One person acts as timekeeper and says 'Go!' The first player picks a name out of the container and has to convey the name to the second player in words or gestures, but without mentioning the name itself. When the second player guesses the name correctly, the first player picks another name from the container and tries to convey this to the second player. The timekeeper stops them after one minute has elapsed. The second player then tries to convey names to the third player, and so on round the circle, until all the pieces of paper have been taken from the container. If players wish to keep scores, they can write down how many names each player guesses in one minute. The game can also be played with other kinds of names or words, such as places, animals, household objects, etc.

Nerbs

23 Two or more players   🖎 Played by speaking

? Punning game .   🖎 No equipment needed

Object To think of ambiguous combinations.

Procedure The players try to say aloud in turn punning phrases or compounds, usually ones in which the second word sounds like a verb even though it may not be a verb. Players have to drop out if they cannot think of a suitable pun.

Example

CHAS Have you ever seen a kitchen sink?
KATE: Have you ever seen a horse fly?
ANNA: Have you ever heard a rubber band?
TONY: Have you ever made a paper weight?
CHAS: Have you ever cashed a health check?
KATE: Have you ever herd of cows?
ANNA: Have you ever felt a hairpin bend?
TONY: Did you know that a tin can?

Never-Ending Story *see* Rigmarole.

Never Say It *see* Taboo.

New Saws *see* Perverbs.

News Reporter *see* How, When, and Where.

N plus Seven *see* Word Substitution.

Nuggets *see* Bullets.

Numericals *see* Logogriphs.

Numwords

②③ Two or more players   ✍ Played by writing
? Word-finding game   ✎ Played with pencils and paper

Object To think of words that add up to a particular number if each letter has a numbered value.

Procedure Each letter of the alphabet is numbered: A=1, B=2, C=3, and so on to Z=26. One player chooses a number, and all the players then write down (within a given time-limit) all the words they can think of whose letters add up to that number. The winner is the person who thinks of the largest number of words. Extra points can be awarded to the person who finds the longest or the shortest word.

Alternatively, players can be asked to think of the highest-value or lowest-value word that fits a particular category: for example, a six-letter word or a word beginning with N.

Example Kate chooses the number 39, and the players have two minutes in which to write down as many words as they can that total 39, if A=1, B=2, etc. Chas writes down *lime, mile, mate, team,* and *hop.* Anna writes *dot, dun, meat,* and *duck.* Tony can only think of *sane* and *sable.* Kate is the winner with *bales, cakes, rat, gages, mile, vice,* and *tar.*

Compare Centurion.

Background Invented by David Parlett.

Nymphabet *see* Wordnim.

O

Odd Man Out *see* Odd One Out.

Odd One Out

⊕⓪⓪ Two or more players	✍ Played by writing or speaking
？ Guessing game	✎ Played with pencils and paper, or with no equipment

Object To guess which one of four items is different from the others.

Procedure One player gives another a list of four words. The other player has to choose the one which has different qualities or characteristics from the other three. So the list might be: *tuba, trumpet, guitar,* and *trombone.* The odd one out is *guitar,* because all the others are wind instruments. Or the list might be: *angelic, devilish, diabolical,* and *hellish,* where *angelic* is the odd one out because all the other words refer to the devil and evil.

Also called Odd Man Out.

O'Grady Says *see* Simon Says.

Oilers

⊕⓪⓪ Two or more players	✍ Played by writing
？ Challenge game	✎ Played with pencils and paper

Object To pick words that share letters.

152

Procedure Oilers is played in two forms. In the first, the players write down these nine words on nine slips of paper:

FISH	SOUP	SWAN
GIRL	HORN	ARMY
KNIT	VOTE	CHAT

Each player in turn takes one of these words. The winner is the first player to take three words that share one letter. If neither player manages to do this, the result is a draw.

In the more complex version of the game, the players write down these 16 words:

APE	DAY	CAN	RAT
LIP	DIE	TIN	RIG
HOP	DOT	ONE	ROW
PUT	BUD	SUN	RUE

Again, each player in turn takes one word. The first player wins if either player takes four words sharing one letter. The second player wins if neither player takes four words that share a letter.

One Awkward Albatross

@@ Two or more players   🖎 Played by speaking
❓ Alphabetical game;   🖌 No equipment needed
     cumulative game

Object To contribute to, and remember, an increasingly long alphabetical list of animals.

Procedure The first player says: 'One awkward albatross.' The second player repeats this and adds a phrase using the letter B, saying something like: 'One awkward albatross, two blue budgerigars.' The third player repeats this and adds an adjective and an animal starting with C, such as: 'One awkward albatross, two blue budgerigars, three cheeky coyotes.' And so on, round the players. Anyone is out who cannot think of a suitable phrase or remember the whole sequence.

Compare Alphabetical Adjectives.

One for Each *see* Guggenheim.

One Hundred Word Challenge *see* One Hundred Words.

One Hundred Words

Any number of players	Played by writing
Challenge game	Played with pencils and paper

Object To write 100 different words that together make sense.

Procedure The challenge of this game is to create a piece of writing that uses 100 different words and makes reasonable sense. This may seem easy, but it is very difficult when you cannot use *and* or *a* or *the* more than once. If several players attempt it, they can set a time-limit and then read one another's attempts.

Also called One Hundred Word Challenge.

One-Letter Omissions *see* Misprints.

One Minute, Please *see* Just a Minute.

One Syllable *see* Monosyllables.

One to One *see* Isosceles Words.

Opposites *see* Antonyms.

Oral Alphabent *see* Alphabent.

Ormonyms *see* Anguish Languish.

Orrible Origins *see* My Word (1).

Outburst!

🎯 Two or more players	✍️ Played on a table with ready-made equipment
❓ Word-finding game	🗃️ Played with a boxed game

Object To think of words which fit particular categories.

Procedure This game is similar to Guggenheim, in that players have to think of words in prearranged categories. The players divide into two teams, and one player is chosen as the MC. The MC picks a 'topic card' from a pack of such cards, which name a topic at the top and list below it ten things which fit that topic. The MC reads out the topic to the first team, which can accept it or pass it to the other team for them to play later. If they pass it, the MC picks out another topic which the first team must then accept. The MC passes the chosen topic card to a player on the second team, who keeps the score. The first team then has one minute to shout out items which they believe will be among the ten listed on the topic card. One point is scored for each correct answer, with bonus points decided by the throw of three dice. The winning team is the first to reach 60 points.

If a team decides to 'pass' a topic, it has to pay one of three 'pass chips' which are given to each team. If one team passes a topic to the other team, that team *has* to play it. If a team uses up all its 'pass chips', it can offer to trade a relatively easy topic for one or more of the other team's chips.

Example The four players divide into two teams: Anna and Chas versus Kate and Tony. They both throw two dice, and Anna and Chas get the higher score, so they will start. Kate acts as MC and picks out a topic card. She tells Anna and Chas that the topic is 'Large Birds'. They pass this topic to the other team, and pay one of their 'pass chips'. Kate picks out another topic and tells Anna and Chas that it is 'Basic Home Tools'. Kate starts the one-minute timer and Anna and Chas shout out: *screwdriver, hammer, pliers, spanner, bradawl, battery charger, tape measure, electric drill, saw, vice, plane,* and *chisel.* Seven of these items are on the topic card, so Anna and Chas score seven points, plus an extra four for *pliers* which has been indicated as a bonus item. Anna and Chas are told that the three items they missed on the topic card are *wrench, wire cutter,* and *spirit level.*

Kate and Tony now have to deal with the topic 'Large Birds'. In one minute they shout out: *vulture, eagle, albatross, ostrich, emu, gannet, pelican, roc, flamingo, stork,* and *peacock.* They score seven points for the seven items which match the topic card, plus a bonus of five for *pelican.* They are told that they missed *hawk, owl,* and *heron.* They don't agree that *owl* is a particularly large bird but they have to abide by the list on the topic card.

Compare Guggenheim.

Background Outburst is a registered trade mark of Hersch & Co.

Oxymorons

Object To find self-contradictory phrases.

Procedure An oxymoron is a phrase which contains words or statements that appear to contradict one another. This includes such phrases as *bitter-sweet*, *pretty ugly*, and *being cruel to be kind*. Oxymorons can be turned into a game in which players take turns to think of contradictory phrases.

Example

ANNA: *Baby elephant.*

KATE: *An open secret.*

CHAS: *Liquid gas.*

TONY: *Light heavyweight.*

ANNA: *Fresh frozen peas.*

KATE: *Accidentally on purpose.*

CHAS: *Plastic glasses.*

TONY: *Military intelligence.*

Compare Irish Bulls.

P

Palindromes

⊕⊘⊘ Any number of players

? Challenge game

✍ Played by writing

✎ Played with pencils and paper

Object To write something that reads the same backwards.

Procedure A palindrome is a word, sentence, or longer piece of writing that spells the same thing if it is read backwards, like *Madam, I'm Adam* or *Able was I ere I saw Elba*. The challenge is to write a palindrome which makes good sense. This becomes harder the longer it is: *Live dirt up a side-track carted is a putrid evil* only just about makes sense.

Palindromes sometimes consist of *words* which are the same if you reverse their order, as in *So patient a doctor to doctor a patient so*. These can be called *pseudodromes*.

Compare Reversals.

Panalphabetic Window *see* Pangrams.

Pangrams

⊕⊘⊘ Any number of players

? Challenge game

✍ Played by writing

✎ Played with pencils and paper

Object To write a sentence containing all the letters of the alphabet.

Procedure A pangram is a sentence that includes all the letters of the alphabet, like *The quick brown fox jumps over the lazy dog.* The challenge is to write as short a pangram as one can, preferably one that includes no letter more than once. It is comparatively easy to devise a sentence like *The five boxing wizards jump quickly,* which repeats only the letters E, I, and U. The more difficult task is to write a sentence containing only the 26 letters of the alphabet.

An associated pastime is to look in printed works for passages which include all the letters of the alphabet. Such a passage is sometimes called a 'panalphabetic window'. For example, a panalphabetic window can be found in Shakespeare's 27th Sonnet.

Pantomime Race *see* The Game.

Paring Pairs

🔵 Two or more players	✍️ Played by writing		
❓ Guessing game	🖊️ Played with pencils and paper		

Object To use clues to find pairs in a list of words.

Procedure One player prepares a number of two-word combinations, and writes cryptic clues to each combination. This player then lists the separate words in alphabetical order. The other player or players then try to find the combinations, using the clues to help them. So the first player might choose the compound *second-hand* and supply the clue 'Left for most people to use'. Combinations can use puns, like *duel personality*, which might be given the clue: 'One or two people who like a fight', or *has bean*, which might be clued as 'Jack was this after selling the cow'. Sometimes an extra word, which is *not* part of the combinations, is included in the wordlist—and the winning player is the one who isolates this odd word by putting together all the other words in pairs.

Example One player prepared this list of clues and the words to be paired to solve them:

Clues

    (1) Where to learn about growing tender plants?
    (2) This figure may bring you friends in time.
    (3) Reach for the sky.
    (4) Swimming baths?
    (5) A motion for a warlike lobster.
    (6) Half the flight requirements for a socialist bird.

Words

A; glass; hour; left; movement; nursery; pincer; pool; room; school; spire; wing.

Solution (1) *nursery school*; (2) *hourglass*; (3) *a spire* (aspire!); (4) *poolroom*; (5) *pincer movement*; (6) *left wing*.

Background Popularized in *Verbatim* magazine.

Parson's Cat *see* Minister's Cat.

Passes *see* Doublets.

Pass It On

⚙ Two or more players
❓ Cumulative game

✍ Played by writing
✎ Played with pencils and paper

Object To build up a story with each player contributing a few words.

Procedure This game is rather similar to Consequences. Each player is given a piece of paper, and writes at the top of it a line and a half to start a story. Each piece of paper is folded over (so that only the second line is visible) and passed to the next player, who completes the second line and starts a third. The papers continue passing from one person to another until each player has made a contribution. The resulting stories are then read aloud. As a variant, a news bulletin can be built up in the same way.

Compare Add a Word; Consequences; Rigmarole.

Password

⚙ Two teams of two, with an umpire
❓ Guessing game

✍ Played by speaking
✎ No equipment needed

Object To guess words from a series of one-word clues.

Procedure The umpire gives one member of each team a 'secret' word written on a piece of paper. The secret word is the same for both teams. The person in the first team who receives the word gives his or her partner a one-word clue to the hidden word. If the partner does not guess it in one try, play passes to the other team, in which the person with the word tries to convey it to his or her partner with a single word. The play continues switching from one team to the other until someone guesses the word. If the word is guessed from the first clue, that team scores ten points; if it is guessed from the second clue, that team scores nine points; and so on downwards. The players from each team can hear what the other team says, thus building up a series of clues to the word.

161

Example Anna and Tony form one team; Kate and Chas form the other. The umpire passes Anna and Kate the word *bucket* written on two separate pieces of paper.

ANNA (*to Tony*): Water.
TONY: Pond?
KATE (*to Chas*): Container.
CHAS: Cistern?
ANNA (*to Tony*): Round.
TONY: Pipe?
KATE (*to Chas*): Handle.
CHAS (*guessing that it is a round water container with a handle*): Bucket?
KATE: Yes!

Also called Keyword.

Patchwork Poetry *see* Centos.

Patchwork Verses *see* Centos.

Pelmanism *see* Associations.

Perverbs

☙ Any number of players

? Challenge game; punning game

✍ Played by writing or speaking

✎ Played with pencils and paper, or with no equipment

Object To create ridiculous proverbs.

Procedure There are four versions of this game, which all involve playing with proverbs.

(1) Players add part of a proverb to part of another proverb to make a comically incongruous saying, such as 'A bird in the hand gathers no moss' or 'When in Rome, do unto others as you wish them to do unto you.'

(2) Players take the first words of a well-known proverb but complete it in a humorous way. For example, 'Where there's a will, there's a large crowd of relatives' or 'He who hesitates is . . . er . . . er . . .'

(3) Many well-known proverbs actually sound rather silly. Players take such proverbs and re-word them so that they make better sense. For example: 'Many hands make light work much more difficult;' 'The grass is just as green on both sides of the fence;' and 'The best form of defence is defence.'

(4) Players make up entirely nonsensical sayings, which nevertheless *sound* as if they could be proverbs—for example, 'Never eat without opening your mouth' or 'Only pick up a blowtorch by the flame if wallpaper-stripping is not your first concern.'

Also called Hack Saws Resharpened; Izzat So?; New Saws; Split Proverbs.

Phrase Maze *see* Get the Message.

Phrases *see* Bullets.

Pi

 Two players
? Grid game

✍ Played by writing
✎ Played with pencils and paper

Object To fill a grid by writing in letters without breaking two rules.

Procedure This game is rather similar to Black Squares. Players alternately write one letter in one square in a grid of blank squares (at least five by five). Any letter can be inserted but three or more adjacent letters (horizontally or vertically) must make a word. And no letter can be used if it prevents the formation of another word of two or more letters. Each player can challenge the other player if he or she is thought to be breaking either of these two rules. If the challenge is correct, the challenged person loses the game. If the challenge is incorrect, the challenger loses the game.

Example Chas starts by writing the letter E in the centre of the grid. Kate adds H above it, and Chas makes the word *hem* by adding M below it. Kate puts a K before the E, and Chas adds an A before the K. Kate almost challenges, suspecting that there is no such word as *ake*, but instead she writes in R before this, making the word *rake*. Chas writes a C before *rake*, which Kate challenges, as she doesn't know such a word as *crake*. But Chas wins, pointing out the word *crake* in the dictionary: meaning a bird.

Pictionary *see* Pictures.

Pictonyms

 Any number of players
? Challenge game

✍ Played by writing or drawing
✎ Played with pencils and paper

Object To write or draw a word in a way that represents its sense.

Procedure Pictonyms are words written or drawn so as to make a kind of picture which represents what the word means. Players can find enjoyment in taking a word from the dictionary at random and trying to 'illustrate' it in the same way as the following examples:

INCOMPLET

exclamation

JAIIIIL

snoozzze

A   LONE

miffionaire

Decembrrr

metricat 10n

HUMiLiTY

OOn
ball

sPOTLIGHT

Also called Qwaints.

Compare Rebus.

Pictures

23 Two or more players (usually two teams)	✍ Played by drawing
? Guessing game	✎ Played with pencils and paper

Object To guess a word or phrase from pictures.

Procedure This is similar to The Game but, instead of the players miming a secret word or phrase, they draw pictures to help their team guess it. One person in each team is told a word or phrase by the referee, and those two players then try to convey it to their teams without speaking but simply by drawing pictures.

Also called (as a boxed game) Pictionary (Copyright 1993 Pictionary Incorporated, Seattle, Washington; distributed in the UK under licence by Hasbro UK Ltd.); (as a television game) Win, Lose, or Draw.

Poetry Consequences *see* Rhyming Consequences.

Poker Crosswords *see* Crossword Game.

Portmanteau *see* Grandmother's Trunk.

Portmanteau Words

⓪②③ Two or more players	✍ Played by speaking
❓ Challenge game	✎ No equipment needed

Object To guess the origins of portmanteau words, or to create new ones.

Procedure A portmanteau word is one that blends together two existing words, like *motel* (a blend of *motor* and *hotel*) or *smog* (a blend of *smoke* and *fog*). Two games can be played with portmanteau words. In the first game, one player thinks of a portmanteau word and asks the next player to say which words are blended to create it. In the second game, players try to make up new, humorous portmanteau words and give their definitions. Thus you might blend the words *hen* and *endurance* to make *hendurance*, meaning 'the patience of a hen trying to hatch out an egg'. Or you could blend the name of the dog *Rin-tin-tin* (who starred in films) and the word *tintinnabulation* to get *Rin-tin-tinnabulation*: a very loud ringing of bells.

Also called Blends.

Preacher's Cat *see* Minister's Cat.

Prefixes

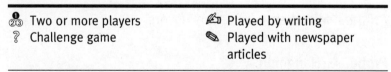

Two or more players

? Challenge game

Played by writing or speaking

Played with pencils and paper, or with no equipment

Object To think of words that start with particular prefixes.

Procedure One of the players suggests a prefix (like *con-* or *multi-* or *semi-*), and all the players think of as many words as they can which start with that prefix. A time-limit can be set, and players can write down their words or speak them in turn, going round in a circle (with anyone dropping out if they cannot think of a suitable word).

Example

ANNA: I suggest we think of words starting with the prefix *anti–*.

KATE: *Antifreeze.*

CHAS: *Antidote.*

TONY: *Antiseptic.*

ANNA: *Anticlimax.*

KATE: *Anticlockwise.*

CHAS: *Antisocial.*

TONY: *Anti . . . er . . . (He has to drop out.)*

Compare Suffixes.

Printers' Errors

Two or more players
? Challenge game

Played by writing
Played with newspaper articles

Object To rearrange a jumbled article in the correct order.

Procedure Each player cuts an article of about a dozen lines from a newspaper or magazine, cuts out the lines separately, and rearranges them—jumbled at random. The jumbled article is given to another player, who has to put it in order. The winner is the first player who restores an article to its original order.

167

Example A player is presented with this jumbled article:

are for hire. Some of them sleep as
The venture was a financial disaster.
many as eight people. The most interesting
However, the canal is now navigable again.
through Oakport Lake and the lock gates,
journey is up the river, through wide
By the time the river reaches Carrick,
anchoring at villages along the way.
gentle tributary, the River Boyle, passing
Drumharlow Lake and into the narrow,
it is deep and sluggish. Houseboats

The successful player rearranges it thus:

The venture was a financial disaster.
However, the canal is now navigable again.
By the time the river reaches Carrick,
it is deep and sluggish. Houseboats
are for hire. Some of them sleep as
many as eight people. The most interesting
journey is up the river, through wide
Drumharlow Lake and into the narrow,
gentle tributary, the River Boyle, passing
through Oakport Lake and the lock gates,
anchoring at villages along the way.

Also called Printers' Mistakes.

Printers' Mistakes *see* Printers' Errors.

Probe *see* Hangman.

Progressive Anagrams *see* Transadditions; Transdeletions.

Progressive Poems *see* Rhyming Consequences.

Prose Poems

⊕ Any number of players

❓ Challenge game

✍ Played by reading and writing

✎ Played with books, plus pencils and paper

Object To find a poem in a piece of prose.

Procedure This pastime consists of looking in prose writings for passages which can be laid out as poems. Occasionally one is lucky enough to find passages that rhyme like poetry, as in Addison's 'What I am going to mention, will perhaps deserve your attention.' More often, one can find passages which can be turned into unrhymed verse, like this passage from prose by Thomas Wolfe:

> And who shall say—
> Whatever disenchantment follows—
> That we ever forget magic,
> Or that we can ever betray,
> On this leaden earth,
> The apple tree, the singing,
> And the gold?

Also called Found Poems.

Proverb Delve

⊕ Any number of players

❓ Word-finding games

✍ Played by writing

✎ Played with pencils and paper

Object To find words in a proverb or other sentence.

Procedure Players choose a proverb or quotation—preferably at random from a dictionary of proverbs or quotations. Within a given time, they try to find as many words as they can in consecutive letters of the sentence, excluding words that are already part of the sentence.

Example Tony opens a dictionary of quotations, closes his eyes, and puts his finger on a page. It lands on the quotation 'I travelled among unknown men, In lands beyond the sea.' He works his way through this quotation, looking for words. He finds *it, rave, ravel, ravelled, led, dam, am, gun, know, inland*, and several others, but he fails to notice other words like *ave, now*, and *me*.

Compare Concealments.

Proverbial Answers

🔢 Two or more players	✍️ Played by speaking
❓ Guessing game	✒️ No equipment needed

Object To solve questions whose answers are in proverbs.

Procedure One player thinks of a proverb and asks the next player (or the other players) a question which can be answered by that proverb. The player who answers correctly can choose to ask the next question.

Example

TONY: What is the mother of invention?

KATE: Necessity is the mother of invention. What should people who live in glass houses *not* do?

TONY: Undress with the light on?

ANNA: No. People who live in glass houses should not throw stones. What is likely to occur when the feline is absent?

CHAS: When the cat's away, the mice will play.

Also called Proverbial Questions.

Proverbial Questions *see* Proverbial Answers.

Proverbs

🔢 Two or more players	✍️ Played by speaking
❓ Guessing game	✒️ No equipment needed

Object To guess a proverb by selecting words from answers given by the other players.

Procedure One person is sent out of the room, while the others choose a proverb. When the outsider returns to the room, he asks each person a question in turn. The first person's answer must include the first word of the proverb, the second person's answer must include the second word, and so on—until the outsider guesses the proverb.

Example Chas goes out of the room, while the others choose the proverb: *Many hands make light work.* Chas returns, and asks Tony: 'Are you really as stupid as you look?' Tony replies: 'That is what many people have told me.' Chas asks Kate: 'How many days are there in a year?' Kate answers: 'I would need more hands than I have got, to be able to count the number on my fingers.' Chas asks Anna: 'Where is the British Museum?' Anna rather unconvincingly replies: 'It is in the Bloomsbury area of London . . . er . . . er . . . make of it what you will!' Chas correctly guesses the proverb.

Also called Guessing Proverbs; Hidden Proverbs.

Pseudodromes *see* Palindromes.

Punchlines

②③ Three or more players   🔊 Played by speaking
? Challenge game   ✎ No equipment needed

Object To end a story with a given line.

Procedure One player becomes the question-master and chooses two 'punchlines' with which two other players must end a story. One player starts telling a story but the question-master interrupts after a short while and the second player then has to continue the story. The question-master switches from one to the other, making it difficult for them to reach their punchlines. The first player to reach a punchline satisfactorily is the winner.

Example Chas is chosen as question-master, and gives Kate the punchline 'And the Hungarian violinists emptied the bath-water on to the chrysanthemums' and he gives Anna the punchline 'So there I was under the table, with an apple-pie on my head.'

KATE: A famous Hungarian orchestra was visiting Britain to play at an agricultural show . . .

ANNA: . . . where I had a stall selling home-made cakes and pies Suddenly . . .

KATE: . . . the orchestra's conductor decided that he needed a bath, so he went into a caravan which had a bathroom, and . . .

ANNA: . . . while he was in there, it looked as if it was going to start raining. I realized that the only place I could shelter from the rain was under the table . . .

KATE: Meanwhile, the conductor finished his bath and came out of the caravan. He noticed that a flowerbed of chrysanthemums was looking rather dry and, as he didn't realize that it was going to start raining, he ordered the musicians to take the water from the bathtub and pour it on the chrysanthemums . . .

ANNA: . . . but it *did* start raining and I crawled under the table to shelter from the rain, unfortunately dislodging an apple-pie which fell on my head. So there I was under the table, with an apple-pie on my head.

CHAS: Well done, Anna!

Also called Tag Wrestling; There I Was.

Pundemonium *see* Tom Swifties.

Puns in Perpetuity *see* My Word (1).

Pyramids (1) *see* Isosceles Words.

Pyramids (2) *see* Word Squares (1).

Q

Quatrains *see* Bouts-rimés.

Question and Answer *see* Questions and Answers.

Questions

⓪ ②③ Two players　　　🖎 Played by speaking
❓ Challenge game　　　🖊 No equipment needed

Object To keep up a conversation using only questions.

Procedure The first player asks a question, and the second player has to reply with another question. The two players have to keep the conversation going, using only questions. No repetitions, grunts, or long pauses are allowed, and the questions must make reasonable sense. A player who cannot think of a question loses one point. If you lose three points, the other player is the winner.

Example This example comes from Tom Stoppard's play *Rosencrantz and Guildenstern are Dead* (published by Faber and Faber in 1967):

GUILDENSTERN:　What's your name?
ROSENCRANTZ:　What's yours?
GUIL:　I asked first.
ROS:　Statement. One-love.
GUIL:　What's your name when you're at home?
ROS:　What's yours?
GUIL:　When I'm at home?
ROS:　Is it different at home?
GUIL:　What home?
ROS:　Haven't you got one?
GUIL:　Why do you ask?
ROS:　What are you driving at?
GUIL:　(*With emphasis*) What's your name?!
ROS:　Repetition. Two-love.

Also called Inquisition; Questions Only, Please; Questions! Questions!

Questions and Answers

 Three or more players   ✍ Played by writing
 ? Cumulative game   ✎ Played with pencils and paper

Object To see what happens when questions are matched randomly with answers.

Procedure Each player is given two slips of paper, and writes down one question and one answer (not necessarily an answer to the same question). The papers are collected and shuffled. Then the questions are read aloud one at a time with an answer to each one, which may be wildly inappropriate. For an alternative version of the game, one of the players prepares a numbered series of questions and answers. The answers only are handed out to the players, who then have to write down what they think the original questions were. The results are read aloud.

Also called Question and Answer.

Questions Only, Please *see* Questions.

Questions! Questions! *see* Questions.

Quizl

 Two players   ✍ Played by writing
 ? Grid game; guessing game   ✎ Played with pencils and paper

Object To guess a word which your opponent has written in a grid.

Procedure Both players draw a grid of five squares by five. They number the columns of the grid from one to five along the top, and from six to nought down the side, so that each square can be identified by a number (e.g. the centre square is 38; the square in the bottom right-hand corner is 50). Each player then writes a five-letter word either across or down in his or her square. They then fill the remaining squares in their grids with 20 other letters which do *not* create any more five-letter words. The players also write on their pieces of paper a second, blank grid of five by five, in which to write their opponent's letters as they discover them.

Each player in turn calls out the number of a square, and their opponent says what letter is in that square. In this way each player can build up a picture of what letters their opponent's grid contains. Instead of guessing a letter, a player can choose to guess the opponent's hidden word: if it is guessed correctly, the guesser wins the game; if not, the game continues. If your opponent's word is guessed correctly, you score one point for every blank square in your opponent's grid. If your opponent has accidentally included more than one five-letter word in the grid, you will win by guessing any of those words.

Example Kate fills in her grid thus (including the word *years* across the middle row):

	1	2	3	4	5	
	S	L	E	E	M	6
	T	A	B	L	I	7
	Y	E	A	R	S	8
	P	U	T	T	E	9
	Q	U	E	A	L	0

Notice that Kate includes several sequences of letters which look as if they might be part of five-letter words but are actually not (like *MISE*, which her opponent might think leads to *miser*, or *QUE* which could be part of several different words). In this way, you can lead your opponent to concentrate on part of the grid which does *not* include the hidden word.

Compare Get the Message; Word Battleships.

Background Invented by David Parlett.

Qwaints *see* Pictonyms.

R

Ragaman

 Two players   ✍ Played by writing
? Grid game   ✎ Played with pencils and paper

Object To build up words within a grid.

Procedure A grid of five by five, seven by seven, or nine by nine blank spaces is drawn. The first player writes a letter in the centre square. The second player has to make a word or anagram by writing a letter in an adjacent square. The players continue alternately, each time adding a letter which makes a word or anagram, horizontally, vertically, or diagonally. Players score points for the number of letters in each new word or words that they make.

Example Anna writes A in the middle square. Chas adds T after it, making the word *at*, and scoring two points. Anna adds F in front of the A, making *fat* and scoring three points. Chas inserts I over the A, making *if* and *it* (diagonally in two different directions) and scoring four points. Anna puts a T diagonally below and to the left of the F, making an anagram of *fit* and scoring three points. As they continue, each player tries to create more than one word or anagram with each new letter, reading in more than one direction.

Compare Word Squares (2).

Railway Carriage Game

⊕ Three or more players   ✍ Played by speaking
? Challenge game   ✎ No equipment needed

Object To introduce a particular sentence into a conversation.

Procedure Two players are chosen to play the game, and a third player gives them each a separate secret sentence. They then have to conduct a conversation and try to introduce their own secret sentence into it before two minutes have elapsed. The first player to introduce his or her sentence is the winner of that round. If the third player judges that one of the players is straying too far from a sensible conversation, that player may be disqualified.

Example Anna writes two sentences on separate pieces of paper. She gives Kate the sentence 'I like it for tea' and Chas the sentence '76 trombones led the big parade.' They start the conversation:

KATE: What is your favourite food?

CHAS: Anything I can eat in the open air.

KATE: Why do you say that?

CHAS: Because I like eating when watching something like a parade.

KATE: I prefer to eat indoors, especially in the afternoon.

CHAS: But there's nothing quite as exciting as being outdoors when a brass band is playing, especially when it plays really loud.

KATE (spotting her opportunity): I like it *forte*. (*Kate may be cheating but the rest of the players declare her the winner of this round, admiring her nerve.*)

Rebus

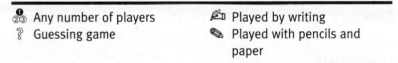

⚇ Any number of players
❓ Guessing game
✍ Played by writing
✎ Played with pencils and paper

Object To create or solve a puzzle in which letters, numbers, symbols, etc. represent words or phrases.

Procedure A rebus uses letters of the alphabet, numbers, pictures, symbols, or other devices to represent a word or words—often a common idiom or saying.

Examples

U 8 PP 4 T   (='You ate peas for tea')

♥ 2 ♥   (='heart to heart')

```
              P
              U
              R
  P   U   R   P   O   S   E
              O
              S
              E          (='cross purposes')
```

Compare Pictonyms.

Background Rebuses are the basis for several games found in newspapers and magazines: e.g. Dingbats and Droodles. Dingbats is also a proprietary boxed game. Catchphrase is a television game-show using rebuses.

Remove a Letter *see* Beheadments.

Reversals

🔟 Two or more players	✍ Played by speaking
❓ Guessing game; word-finding game	✎ No equipment needed

Object To find or guess words which make other words backwards.

Procedure Many words make other words when they are read backwards, such as *draw* (which becomes *ward*) and *loop* (which becomes *pool*). Two games can be played with such words. Players can take it in turns to think of such pairs of words—and any player is out of the game if they cannot think of a suitable pair. Or one player can give the other players clues to such pairs of words for them to guess. If a player guesses the words, that player chooses the next pair of words to be reversed.

Example

ANNA: Can you reverse the action of swallowing to make a wedge or stopper?

KATE: *Plug* and *gulp*. Can you reverse a section to make an ambush?

CHAS: *Part* and *trap*. Can you reverse puddings to get a word that means 'under pressure'?

ANNA: *Desserts* and *stressed*.

Also called Backwords; Bacronyms; Semordnilaps.

Compare Palindromes.

Rhopalics

②③ Any number of players

? Cumulative game

✍ Played by writing or speaking

✎ Played with pencils and paper, or with no equipment

Object To construct a sentence, etc. in which each word has one more syllable than the preceding word.

Procedure The aim of rhopalics is to construct a sentence or even a poem consisting of words which have a progressively larger number of syllables, so that the first word has one syllable, the second word has two syllables, and so on. This can be tried by a person alone or by several people each contributing a word in turn. As it is difficult to write sensible rhopalics of more than about seven or eight words, the cycle can be repeated after (say) every fifth word. If a poem is constructed, every new line can start afresh with a one-syllable word. An alternative, easier form of the rhopalic consists of words which each have one *letter* more than the preceding word.

Example Hope always solaces miserable individuals inconspicuously.
Also called Rhopalic Verses; Snowball Sentences.

Rhopalic Verses *see* Rhopalics.

Rhymed Endings *see* Bouts-rimés.

Rhyme in Time

ⓞ Two or more players	✍ Played by speaking
? Challenge game; rhyming game	✎ No equipment needed

Object To keep up a spoken rhyming sequence.

Procedure The first player says aloud a phrase, and subsequent players in turn must say a different phrase that rhymes with it. Anyone who cannot think of a rhyming phrase has to drop out. Players can also be disqualified if their phrase does not follow on naturally from the preceding phrase, or if they repeat a rhyming word that has already been used.

Example

KATE: What's for tea?
ANNA: Don't ask me.
CHAS: Is it free?
TONY: We shall see.
KATE: I agree!
ANNA: Is there jell-ee?
CHAS: And nice pastry?
TONY: That's for me! (*He has to drop out, as he has repeated a rhyme that has already been used.*)

Rhyming Consequences

23 Two or more players
? Cumulative game;
rhyming game

🖎 Played by writing
✎ Played with pencils and
paper

Object To build up a ridiculous series of verses.

Procedure This is a variation of the game Consequences. In this version, each player writes down two rhyming lines, and the first half of a third line. The players then fold over the paper so that only the half-line can be seen, and pass it on to the next player, who completes that line and adds another rhyming line and an uncompleted line. When the papers have passed from one player to another several times, the players read out the 'poem' they are holding. Alternatively, players can write just one line each, with the final word shown so that the next player can add the next rhyming line.

Example Tony writes down these lines:

> I think that I shall never see
> A person quite as beautiful as me.
> For I'm a very ...

He folds over the paper to conceal the first two lines, and passes it to Kate, who continues:

> For I'm a very wicked lady
> With manners poor and habits shady,
> And when I go ...

She passes this on to Chas, who continues:

> And when I go to bed at night
> I always eat Turkish delight.
> That's why they call me ...

Anna completes the poem:

> That's why they call me Gunga Din
> And always try to lock me in.

Also called: Poetry Consequences; Progressive Poems.

Rhyming Ends *see* Bouts-rimés.

Rhyming Tom *see* Crambo.

Riddle-Me-Ree

🎲 Any number of players	✍ Played by writing
❓ Guessing game	✎ Played with a previously prepared puzzle, plus pencils and paper

Object To build up the letters of a word from other words which include or exclude those letters.

Procedure A riddle-me-ree is a puzzle in verse form, usually listing alternately words that contain and do not contain each letter of the mystery word.

Example

My first is in *dog* but not in *cat*,
My second is in *bowler* but not in *hat*,
My third is in *wave* but not in *ripple*,
My fourth is in *drink* but not in *tipple*,
My last is in *piece* but not in *fraction*,
And my whole is a famous man of action.

This riddle-me-ree was written by Hubert Phillips. The answer is: *Drake*.

Riddles

🎲 Any number of players	✍ Played by writing or speaking
❓ Guessing game	✎ Played with pencils and paper, or with no equipment

Object To guess the answer to a puzzling question or statement.

Procedure A riddle is a question or statement designed to puzzle people, often by phrasing the question or statement in a deliberately mysterious or misleading way: for example, 'Why does the Archbishop of Canterbury wear red, white, and blue braces?' (Answer: To hold his trousers up). If the riddle depends on a pun, it is a *conundrum:* for example, 'When is a door not a door?' (Answer: When it's *ajar*).

Compare Enigma.

Rigmarole

⓪②③ Two or more players	✍ Played by speaking
? Cumulative game	✎ No equipment needed

Object To build up a story.

Procedure One player starts telling a story, but stops after one or two sentences (or in the middle of a sentence). The next player has to continue the story, again stopping at some point for it to be continued by the next player. To make the game competitive, it can be ruled that any player who *finishes* the story has to drop out of the next round.

Also called Chain Story; Group Story; Never-Ending Story; Serial Story.

Compare Add a Word; Consequences; Pass It On.

Roman Numeral Words *see* Chronograms.

Russian Gossip *see* Chinese Whispers.

Russian Scandal *see* Chinese Whispers.

S

Sausages

👥 Two or more players   📢 Played by speaking
❓ Challenge game   ✎ No equipment needed

Object To avoid smiling when answering questions with a ridiculous word.

Procedure One or more of the players ask one or more of the others a series of questions in turn. The answers have to consist simply of the word *sausages*. If the answerer laughs or smiles, he or she is out of the game. Other ridiculous words can be used instead of *sausages*: for example, *bananas, boloney,* or *knickers.*

Example Tony is chosen as the person to be questioned.

KATE (*to Tony*): What is your name?

TONY: Sausages.

ANNA: Which flowers look nice in the garden?

TONY (*almost smiling*): Sausages.

CHAS: What do your fingers look like?

TONY: Saus... (*Tony collapses with laughter.*)

Also called Bananas; Boloney; Knickers.

Scattergories   *see* Guggenheim.

Scorewords   *see* Crossword Game.

Scrabble

👥 Two, three, or four players   📢 Played on a board
❓ Grid game   ✎ Played with a boxed set of a board and 100 letter-tiles

Object To form words on a grid.

Procedure Scrabble uses a board of 225 squares (15 by 15), on which players form interlocking words using 100 'tiles' bearing letters of the alphabet. Each letter has a score beside it. The board includes various 'premium squares' on which players can score extra points.

The number of letters in a Scrabble set and their scores are as follows:

Letter	Number of tiles	Value	Letter	Number of tiles	Value
A	9	1	O	8	1
B	2	3	P	2	3
C	2	3	Q	1	10
D	4	2	R	6	1
E	12	1	S	4	1
F	2	4	T	6	1
G	3	2	U	4	1
H	2	4	V	2	4
I	9	1	W	2	4
J	1	8	X	1	8
K	1	5	Y	2	4
L	4	1	Z	1	10
M	2	3	Blank	2	0
N	6	1			

All the letter tiles are turned face downwards or kept in a bag. To decide who will start the game, each player picks out one tile and the person drawing the letter nearest to 'A' plays first. Each player picks seven tiles and places them on a ready-made rack, where only that player can see them. The first player puts letters on the board to make a word, either 'across' or 'down' (as in a crossword), with one letter of the word on the central square. The score is recorded, including any double or triple allowance for tiles placed on premium squares. The first word always scores double, as the centre square has a 'double word score'. A bonus of 50 points is awarded to any player who uses all seven tiles in one move. The first player picks out from the unused tiles the same number of tiles as was used to make the word, so as to make up the number to seven again.

The next player then has to add another word, joining or interlocking with the word on the board, and so on—round the players. All new words must use at least one of the letters that is already on the board. Players score for any word made or changed by their moves—but premium bonuses apply only the first time that letters are played. Instead of laying down a word, any player can exchange any number of tiles from their rack for new tiles from the 'bank'. The Scrabble set includes two blank tiles, which can represent any letter its player chooses, after which it cannot be changed during the game.

The game ends when all the tiles have been used and one player has laid down all his or her tiles, or nobody can think of new words to place on the board. The players have to deduct from their scores the value of all their unplayed letters—and this total is added to the score of a player who has disposed of all his own tiles. The winner is the player with the highest score.

There are several variations of Scrabble. Double-Bag Scrabble divides the tiles into two separate bags: one for the vowels, the other for the consonants. Scrabble for Juniors is a boxed game using a two-sided board: on one side, children can play a normal game of Scrabble except that there are only 13 by 13 squares and the scoring is simplified; on the other side, young children have to place letter tiles on words already printed on the board. Solitaire Scrabble is for one player only. Unscrabble (or Scrabble in Reverse) involves removing letters from the board: players remove between one and six tiles from the board at each move, but the letters they leave must spell proper words interlocked with one another. This game ends when nobody can remove any more tiles from the board, and the winner is the player whose stock of removed tiles makes the highest score.

Example Each player picks a tile from the bag, and Kate starts the game because she picks the letter nearest to the start of the alphabet.

Kate's first seven letters are A, C, E, G, I, N, and N. She wants to use the longest word she can find with the highest-scoring letters, so she chooses *caning*, which she puts across the centre of the board so that the C is on a 'double letter score' square. Thus the word scores 12 but Kate gets 24 points as the word is on the double-scoring centre square.

Anna draws the seven letters E, H, I, L, N, R, and U. She keeps the U in case she later draws a Q (as many words start with *qu–*) and uses the other letters vertically with the A in *caning* to spell the word *inhaler*. She positions it deliberately to use two 'double word score' squares, so that the basic score of ten is quadrupled to 40.

Chas draws the letters A, E, I, O, T, V, and Y. He makes the word *veiny* across the N of *inhaler*, so that the high-scoring letters V and Y are both on 'triple letter score' squares, thus scoring a total of 27.

Tony draws the letters A, C, D, E, R, T, and T. He makes the word *rate* horizontally, starting with the R from *inhaler*, not seeing that he could make longer words like *rated*, *crater*, or even *retract* (which would reach a 'double word score'). The board now looks like this:

```
        I
V   E   I   N   Y
        H
    C   A   N   I   N   G
        L
        E
        R   A   T   E
```

Kate draws the letters D, H, K, M, S, and W to add to her remaining E. She holds on to the letter S, which is always useful to add at the end of words (as are such letters as D, E, N, and R). She puts the H to the right of the E of *inhaler* and above the A of *rate*, making two words—*eh* and *ha*—and scoring 26 with the help of the triple letter score. She has remembered the importance of trying to make more than one word with each move, and she has also memorized a number of acceptable two-letter words, since these are often useful in Scrabble. Short words can score more than long words: for example, *zoo* scores 12 but *strain* scores only six.

Anna draws the letters B, E, E, Q, V, and Z. She now has the Q to go with her U, so she makes *queen* ending with the second N of *caning*. This scores 24 because the Q is on a 'double letter score'.

Chas draws the letters D, I, J, and T to add to his A, O, and T. He sees that he can make *titrate* (a word in chemistry) by adding to the word *rate* which is already on the board. However, before playing this word, he wisely looks to see if he can add letters at both ends of *rate*, and discovers that he can make *titrated*.

Tony draws A, F, and a blank tile to add to his C, D, R, and T. He makes the word *tread*, running vertically from the last T of *titrated* (and using the blank tile as an E) but he would have scored more if he had put the same letters under the first or second T of *titrated*, as this would have given him a double word score. The board now looks like this:

```
                        Q

            I           U

V   E   I   N   Y       E

            H           E

        C   A   N   I   N   G

        L

        E   H

T   I   T   R   A   T   E   D

            R

            □

            A

            D
```

Clearly Tony might benefit from learning more about the tactics of Scrabble. As well as the points mentioned above, he could try to: (1) avoid giving opportunities to his opponents, especially in opening up to them the use of the premium squares; (2) do his best to find seven-letter words (so as to score the 50-point bonuses); (3) learn words that use high-scoring letters like J, Q, X, and Z, and try to get these letters on to the premium squares; (4) remember which tiles have already been used (so as to assess the likelihood of making particular words); (5) save the blank tiles until they can be really useful in a high-scoring word; (6) look for opportunities to benefit from his opponents' high-scoring words by adding letters to them; and (7) exchange his unplayable tiles so that his opponents may pick them, especially towards the end of a game.

Background Scrabble is the registered trade mark of J. W. Spear & Sons plc.

Scrabble for Juniors *see* Scrabble.

Scrabble in Reverse *see* Scrabble.

Scramble

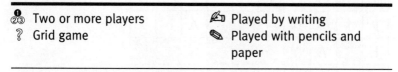

⚅ Two or more players		✍ Played by writing	
? Grid game		✎ Played with pencils and paper	

Object To fill a grid with words connected to one theme.

Procedure Each player draws a grid of (say) ten squares by ten. The players agree on a subject, such as sport, natural history, art, or food. The players then try to fill their grids with words connected with the chosen theme, making the words interlock (as in a crossword) and blacking out any unused squares. As soon as one player has filled his or her grid, this player shouts out and everyone has to stop writing. The player who shouts out must have put letters in at least half the squares in the grid. Players pass their grids to the player on their left for checking. One point is scored for each letter in every acceptable word. The winner is the person with the most points.

Example The chosen theme is 'Animals' and Chas fills his grid like this:

C	H	I	M	P	A	N	Z	E	E
H			A		P				L
I	B	I	S		E				E
N			T				P	U	P
C	O	L	O	B	U	S			H
H			D				C		A
I		G	O	P	H	E	R		N
L	I	O	N				A	N	T
L		A					N		
A	N	T	E	L	O	P	E		

For this, Chas scores 83 points (45 for the letters in the words 'across'; 38 for the words 'down').

Compare Crossword Game.

Secret Word *see* Jotto.

Semantic Poetry

 Any number of players

? Wordplay game

✍ Played by writing

✎ Played with pencils and paper, and a dictionary

Object To replace words in a piece of writing with their dictionary definitions.

Procedure The point of this game is to see what happens when one substitutes dictionary definitions for some or all of the words in a piece of poetry or prose. For example, by replacing words in the first lines of Shakespeare's *Macbeth* with their definitions from the *Concise Oxford Dictionary*, one can change: 'When shall we three meet again: In thunder, lightning, or in rain?' into:

At what time shall we one more than two join or fasten correctly another time:
In a loud rumbling or crashing noise heard after a lightning flash, a flash of bright light produced by an electric discharge, or in the condensed moisture of the atmosphere falling visibly in separate drops?

Compare Word Substitution.

Semordnilaps *see* Reversals.

Sentence Building *see* Telegrams.

Sentences *see* Short Story; Telegrams.

Sequences

⚛ Any number of players   ✍ Played by writing
? Challenge game   ✎ Played with pencils and paper

Object To find words containing consecutive letters of the alphabet.

Procedure Within a given time, players have to write down as many words as they can think of that include two consecutive letters of the alphabet side by side: for example, *able* (which has A next to B) or *fastidious* (which contains S and T next to one another). If this is too easy, players can do the same with sequences of three, four, or even more letters. For example, the word *understudy* includes the sequence R–S–T–U.

Also called Sequentials.

Sequentials *see* Sequences.

Serial Story *see* Rigmarole.

Short Story

⚛ Any number of players   ✍ Played by writing
? Challenge game   ✎ Played with pencils and paper

Object To write something that uses only short words.

Procedure The challenge of this game is to write a story using very short words. Players have to use monosyllables, or words of only three letters, or no word longer than three letters. If the game is played competitively, the winner is the player who writes the longest story.

Example Ann and Jim saw the red cat on the bed. Kit is a big cat and yet not so big as to be as big as a log. Now the cat can get a nap and . . .

Also called Sentences.

Compare Monosyllables.

Shouting Proverbs

Ⓥ Four or more players   ✍ Played by speaking
? Guessing game   ✎ No equipment needed

Object To guess a proverb when other players shout separate words simultaneously.

Procedure The players divide into two teams. One team goes out of the room and chooses a proverb. Each member of this team takes one word of the proverb. For example, if the proverb is *A rolling stone gathers no moss*, the first player takes *A*, the second takes *rolling*, and so on. If there are more players in the team than there are words in the proverb, two people take the same word. This team then returns to the room and, facing the other team, shouts out their words all at the same time. The other team has to identify the proverb: if it fails, the challengers go outside again and choose another proverb to shout. Other phrases instead of proverbs can be used: for example, titles of books or films. The game can also be played with all the players except one shouting a proverb at that one person.

Also called Simultaneous Proverbs; What Did We Say?

Shrink Words

Ⓥ Any number of players   ✍ Played by writing
? Challenge game   ✎ Played with pencils and paper

Object To reduce a word to one letter by deleting or substituting letters.

Procedure Players choose a word to start with and they try to change it by stages into a single-letter word, either by deleting a letter or by substituting one letter for another. A new word must be created at every step of the game.

Example The player chooses the word *thanks*. This can be changed into a one-letter word in this way:

THANKS

THANK (delete the S)

THINK (change A to I)

THING (change K to G)

TING (delete the H)

SING (change T to S)

SIN (delete the G)

IN (delete the S)

I (delete the N)

Compare Beheadments; Doublets; Isosceles Words; Transdeletions.

Side by Side

 Two or more players   ✍ Played by speaking

❓ Guessing game   ✎ Played with a dictionary

Object To guess words which are next to one another in the dictionary.

Procedure One of the players takes a dictionary and picks out a pair of words which are next to each other. This player reads out the two definitions to the other players, who try to guess the pair of words. The person who guesses correctly is allowed to choose the next pair of words.

Example Anna chooses a pair of words from the dictionary, and asks the others: 'What are these two words, which are next to one another in the *Concise Oxford Dictionary*? One means "a circular spangle for attaching to clothing as an ornament" and the other means "a Californian evergreen coniferous tree".'

Chas thinks the second word might be *redwood* but he cannot think of a word for a spangle which would precede *redwood* in alphabetical order. Tony is puzzled as he thinks a *spangle* is a kind of sweet, and he cannot think why it would be attached to clothing. Kate guesses that the first word is *sequin*, which leads her to think that the second word probably starts with *seq*–, and eventually she guesses correctly that the two words are *sequin* and *sequoia*.

Compare Dictionary Game.

Simon Says

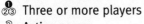

⓵⓶⓷ Three or more players	✍ Played by speaking
❓ Active game; challenge game	✎ No equipment needed

Object To follow instructions according to the way they are phrased.

Procedure One player becomes 'Simon' and gives the other players instructions to perform particular actions. The players must only follow his instructions if they are preceded by the words 'Simon says'. So, if the leader says: 'Simon says, touch your toes,' the players must touch their toes. But if the leader simply says: 'Touch your toes,' anyone who obeys is out of the game. The last one remaining is the winner, and becomes the next leader.

Alternative versions of the game are called O'Grady Says or Do This, Do That' (in which players must obey orders preceded by 'Do this' but ignore orders preceded by 'Do that').

Compare Sophie Says.

Simplets *see* Bullets.

Simultaneous Proverbs *see* Shouting Proverbs.

Sinko

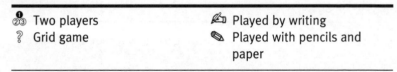

Two players	Played by writing
Grid game	Played with pencils and paper

Object To fill a grid with five-letter words.

Procedure A grid of five squares by five is drawn. The first player writes a five-letter word anywhere in the grid, horizontally or vertically. The second player then has to insert another five-letter word, either parallel to the first word or interlocking with it. The players continue alternately writing in words until neither player can fill in another word. The winner is the player who inserts the last word.

Example Chas starts by writing *treat* down the centre of the grid. Kate inserts *batty*, intersecting the first T of *treat*. Chas enters *years*; Kate inserts *brawl*; and Chas enters *lutes*, so that the grid looks like this:

B	A	T	T	Y
R		R		E
A		E		A
W		A		R
L	U	T	E	S

Chas thinks he has won, as he cannot see any word that would fit into the grid, but Kate sees that she can insert *ameba* (an American spelling of *amoeba*) across the middle of the grid, so she wins.

Snowball Sentences *see* Rhopalics.

Solitaire Scrabble *see* Scrabble.

Sophie Says

🔢 Three or more players	✍ Played by speaking
❓ Active game; challenge game	✎ No equipment needed

Object To follow instructions according to the way they are phrased.

Procedure This game is similar to Simon Says. The players form a circle, with a 'leader' in the centre. The leader gives instructions for actions (like 'Put your hands on your head') but the players must obey these orders only when they are preceded by the words 'Sophie says'. Anyone who obeys an order that is *not* preceded by these words has to start forming a circle inside the original circle. In the end, everyone is in the inner circle.

Spellbound *see* Backward Spelling Bee.

Spelling Bee

👥 Two or more players	✍ Played by writing or speaking
❓ Spelling game	✎ Played with a dictionary

Object To compete in spelling words correctly

Procedure One player is chosen as the question-master and chooses words from a dictionary for each of the other players to spell. Points are scored for spelling words correctly. If someone spells a word incorrectly, either that person drops out of the game, or the chance passes to the next player. The game may also be played in two teams, and the team members can choose words for the other team to spell. Alternatively, a list of (say) ten words can be read out, which the players have to write down correctly.

Also called Spelling Quiz.

Spelling Crab *see* Backward Spelling Bee.

Spelling Quiz *see* Spelling Bee.

Spelling Round

👥 Three or more players	✍ Played by speaking
❓ Spelling game	✎ Played with a dictionary

Object To spell words with each player contributing one letter in succession.

Procedure The players sit in a circle. The question-master chooses a word for the players to spell, and points to one of the players in the circle. That player gives the first letter of the word, the next player adds the second letter, and so on round the circle until the word is complete. Anyone making a mistake or hesitating is out of the game. The winner is the last player to survive.

Split Proverbs *see* Perverbs.

S plus Seven *see* Word Substitution.

Spoken Charades *see* Charades.

Spoonergrams *see* Spoonerisms.

Spoonerisms

😀	Any number of players	✍	Played by writing or speaking
❓	Wordplay game	✎	Played with pencils and paper, or with no equipment

Object To use phrases which interchange parts of words.

Procedure Spoonerisms are often caused unconsciously, when a speaker absent-mindedly transposes one or more letters (usually the first letter or letters) of adjacent words, saying 'it's roaring with pain' when he means 'it's pouring with rain', or 'a blushing crow' instead of 'a crushing blow'. People can gain amusement from creating their own spoonerisms and sharing them with other people. Spoonerisms can also be turned into a game where one person asks another to solve a clue which leads to a spoonerized phrase. For example, when spoonerized, which part of the foot might describe Manx cats? The answer is *toe nails*, which can be spoonerized into *no tails*.

Another game with spoonerisms is Spoonergrams, which are poems or other pieces of writing in which blank spaces are to be filled by spoonerisms. For example:

On summer days I love to see the ——
And watch it, brightly-coloured, —— ——.

The blanks can be filled with *butterfly* and *flutter by*.

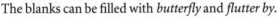

201

Background Named after the Revd W. A. Spooner, Warden of New College, Oxford, from 1903 to 1924, who was supposed to use these unconscious transpositions.

Squared Words *see* Crossword Game.

Staircase *see* Stairway.

Stairway

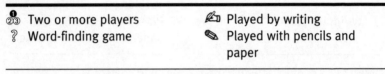

Two or more players	Played by writing
Word-finding game	Played with pencils and paper

Object To build up a 'stairway' of words of ever-increasing length.

Procedure One player chooses a letter of the alphabet, and all the players have to write down (within a set time) words beginning with that letter: a two-letter word, a three-letter word, and so on with longer and longer words. There must be no gaps in the 'stairway'. The winner is the person who builds up the highest stairway in the given time.

Example Within the given time-limit of five minutes, Anna built up
this stairway of words beginning with S:

S
SO
SET
SUIT
SWEET
SUNSET
SEAWEED
SUSPENSE
SEBACEOUS
SMATTERING
SYMPTOMATIC
SALPIGLOSSIS
SCANDALMONGER
SUSCEPTIBILITY
SCINTILLATINGLY
SESQUICENTENNIAL

Also called Staircase.

Stars *see* Word Squares (1).

Starters

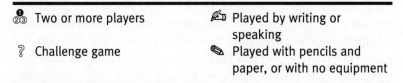

🎲 Two or more players		📖 Played by writing or speaking	
⁇ Challenge game		✏ Played with pencils and paper, or with no equipment	

Object To think of words which start with particular letters.

Procedure This game can be played in two ways. Players can take turns to ask other players to think of a word that starts with a particular series of letters. Alternatively, players can be given a list of letter-groups and challenged to write down one word that starts with each group. The winner is the first player to write down suitable words for the whole list.

Example

KATE: Can you think of a word that starts with LAN–?

TONY: *Language.*

ANNA: *Languid.*

CHAS: *Langoustine.* Can you give me a word that starts with PUR–?

KATE: *Purple.*

TONY: *Purpose.*

ANNA: *Purchase.* Can you think of a word that starts with TUF–?

CHAS: *Tuft.*

KATE: *Tufa.*

TONY: *Tuffet.*

Stepping Stones

@③ Two or more players

🖎 Played by writing or speaking

? Challenge game

✎ Played with pencils and paper, or with no equipment

Object To move from one word to another by way of specific categories.

Procedure One player suggests to the next player two words or subjects that must be connected by a string of other words, including on the way two or three particular words or subjects. The challenged player has to create a series of steps in which each word has some association with the next (compare Associations). All the other players can decide if the connections between one word and the next are acceptable. If more than two people play, they can all try to make the connection, and read out their answers. If played competitively, the game is won by the player who completes the connection in the smallest number of steps.

Example Chas challenges Kate to move from *cars* to *ducks*, by way of *fish-and-chips*, *gardens*, and *electricity*. Kate achieves it thus:

Cars need *keys* to start.

This sounds like *quays*, where *boats* are tied.

Boats are used to catch *fish* (often served with *chips*).

Fish are sometimes kept in *bowls*.

Bowls is a game played on *grass*.

Grass is found in many *gardens*.

In *gardens* people plant *bulbs*.

Bulbs are used in *electric* lights.

Electricity can give you a *shock*.

You may get a *shock* when you receive a *bill*.

A *bill* is another name for the beak of a *duck*.

Also called Explain That.

Compare Associations; Tennis, Elbow, Foot.

Stepwords *see* Doublets.

Stinkety Pinkety *see* Stinky Pinky.

Stink Pink *see* Stinky Pinky.

Stinky Pinky

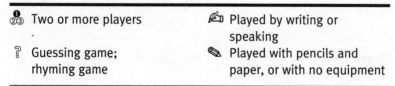

| ⓪②③ Two or more players | ✍ Played by writing or speaking |
| ? Guessing game; rhyming game | ✎ Played with pencils and paper, or with no equipment |

Object To think of rhyming phrases to fit definitions.

Procedure One player thinks of a phrase of two rhyming words (usually an adjective and a verb), and then tells the other players a definition which would lead to that rhyming phrase. The first player to think of a suitable phrase can then propose another definition for the other players to guess. The game can also be played by two players, swopping 'stinky pinkies'. In some forms of the game, a rhyming phrase of two monosyllables is called a *stink pink*; a phrase with two-syllable words is a *stinky pinky*; and a phrase of three-syllable words is a *stinkety pinkety*.

Example

KATE: Here's a stink pink. What is a two-word rhyming phrase for a large hog?

CHAS: A *big pig*. What would be a stinky pinky for 'a rabbit that makes you laugh'?

ANNA: *Funny bunny*. Here's a stinkety pinkety: 'a hatred of entertainments'.

TONY: *Diversion aversion*. How about 'clergyman's underpants'?

KATE: *Vicar's knickers!*

Also called Hankety Pankety; Hank Pank; Hanky Panky; Hinkety Pinkety; Hink Pink; Hinky Pinky; Stinkety Pinkety; Stink Pink.

Stray Syllables *see* Syllables.

Suffixes

② Two or more players	✍ Played by writing or speaking
? Challenge game	✎ Played with pencils and paper, or with no equipment

Object To think of words that end with particular suffixes.

Procedure One of the players suggests a suffix (like *–ance* or *–ment*), and all the players think of as many words as they can which end with that suffix. A time-limit can be set, and players can write down their words or speak them in turn, going round in a circle (with anyone dropping out if he or she cannot think of a suitable word).

Example

CHAS: I suggest we think of words ending with the suffix *–logy*.
KATE: *Biology.*
ANNA: *Archaeology.*
TONY: *Philology.*
CHAS: *Zoology.*
KATE: *Trilogy.*
ANNA: *Speleology.*
TONY: Er . . . I can't think of any more. (*He has to drop out.*)

Compare Prefixes.

Suggestion Chain *see* Associations.

Superghost *see* Superghosts.

Superghosts

 Two or more players
? Cumulative game;
letters game;
word-building game

✎ Played by speaking
✎ No equipment needed

Procedure A variation of *Ghosts*, played in the same way except that letters can be added at the beginning of a word as well as at the end.

Example Chas says F, thinking of *flag*. Kate makes it FL, thinking of *flexible*. Anna makes it IFL, thinking of *stifle*. Tony makes it RIFL, thinking of *rifle*. Chas has to complete it, because he cannot think of any possible word except *rifle*, but he could have said TRIFL (thinking of *trifle*) or RIFLI (thinking of *rifling*) to avoid losing a 'life'.

Also called Superghost.

Syllables

②③ Two or more players   📖 Played on a table
? Word-building game   ✎ Played with pencils and paper

Object To make words from syllables

Procedure Each player writes on strips of paper a number of words, with gaps between their syllables. It does not matter if the words are not split up according to the strict rules of word-division. The individual syllables are then cut up and shuffled into a pile. Each player in turn draws from the pile three pieces of paper, and tries to form a word on the table from one or more of the syllables. If players cannot make a word, they return their slips of paper to the pile and draw three more at their next turn. When all the slips of paper have been used, and nobody can make any more words, the winner is the person who has made the most words.

Example Each player writes down four words. Anna writes *dis-miss*, *de-pend-ent*, *im-port-ance*, and *dog-fish*. Kate writes *trick-ling*, *e-pis-co-pal*, *in-ter-est*, and *for-ti-fi-ca-tion*. Chas writes *ger-bil*, *ap-pear-ance*, *min-is-ter*, and *in-ter-na-tion-al*. Tony writes *de-pen-dent*, *prob-ab-ly*, *cha-rac-ter-ist-ic*, and *fin-ic-ky*. The words are cut up into syllables, shuffled, and placed in a pile in the middle of the table. Anna starts by picking out the syllables *tion*, *ter*, and *na*. She makes the word *nation* and returns *ter* to the pile. Kate draws *pend*, *bil*, and *pear*. She cannot see a word of two or three syllables, so she lays down *pear* and discards the other two slips of paper. Chas draws *im*, *min*, and *ance*, and he makes *minim*. Tony draws *is*, *cha*, and *in*, and makes *chain*. Anna draws *ger*, *fin*, and *fish*, and makes *fish-finger*. Kate draws *for*, *de*, and *est*, and makes *deforest*. Chas draws *bil*, *ky*, and *ly*, and makes *billy*. Tony draws *dis*, *pal*, and *trick*, and tries to make a word *dis-trick* but the other players point out that it should be *district*. Play continues and the players make more words, including *pendent*, *dental*, *missis*, *disco*, *probe*, *deist*, *cater*, *inter*, and *penance*.

Also called Stray Syllables.

Synonym Chains

Any number of players

? Challenge game; cumulative game; word-finding game

Played by writing or speaking

Played with pencils and paper, or with no equipment

Object To build up a chain of synonyms leading from one chosen word to another.

Procedure Any player proposes a pair of words, preferably two that mean the opposite of one another (such as *true* and *false*). The object is to think of a synonym for the first word (i.e. a word that means approximately the same), and then a synonym of *that* word, and so on, until the chain leads to the second chosen word. If there are any disagreements about the acceptability of some synonyms, they can be checked in a reputable dictionary of synonyms.

Example The two words chosen are *light* and *dark*. One way of getting from one to the other by means of synonyms is: *light–bright–clever–cunning– sly–furtive–hidden–obscure–dark*.

Synonyms

Two or more players
? Word-finding game

Played by writing

Played with a previously prepared list of words, plus pencils and paper.

Object To find synonyms for given words.

Procedure A list of words (between about ten and twenty) is prepared, and a copy is given to each player. Within a given time-limit (say, five minutes) the players try to write alongside each word a synonym for it — that is, a word with the same meaning. If players can think of more than one synonym, they can write down as many as they can find. The winner is the player who finds the highest number of acceptable synonyms. If there are any queries about acceptability, players can refer to a dictionary of synonyms to decide which words can be accepted.

Example Each player is given a copy of the list of chosen words. The list contains the words *help, calm, tooth, feeble, speech, annoy, sweetheart, sharp, peace,* and *job.* For *help,* all the players write down *assist* and *aid,* but Kate also thinks of *succour* and *support,* while Chas thinks of *serve* (as in 'help yourself to food'), *backing,* and *co-operation.*

Syzygies

⓪②③ Any number of players	✍ Played by writing
？ Challenge game	✎ Played with pencils and paper

Object To change one word into another in a number of stages.

Procedure Two words are chosen, and the aim is to change one word into the other by moving through words which have a sequence of some of the same letters. At each stage, some of the letters in *one* word occur in the same order in the *following* word.

Example You can change *atom* into *bomb* by taking the TOM of *atom* and finding another word that contains that sequence of letters: *tomb.* Then simply take the OMB of *tomb* and it is part of *bomb*:

A   T   O   M

     T   O   M   B

     B   O   M   B

For a more complicated change, you can make *Wednesday* into *afternoon* in five steps. Starting with *Wednesday,* take out the letters EDNES and think of another word that includes them: *blessedness.* From this word, take the letters ESSE, which also occur in *finesse.* Take out INESS, which also appears in *craftiness.* Take out RAFT, which is also in *rafter,* from which AFTER leads you to *afternoon*:

```
        W E D N E S D A Y
            E D N E S
    B L E S S E D N E S S
            E S S E
      F I N E S S E
          I N E S S
C R A F T I N E S S
R A F T
R A F T E R
A F T E R N O O N
```

Background Game invented by Lewis Carroll (compare Doublets). Its name is pronounced 'siz-i-jiz' (Carroll said it rhymed with 'fizzy fizz').

T

Taboo

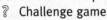 Two or more players
? Challenge game

✍ Played by speaking
✎ No equipment needed

Object To avoid saying a chosen word.

Procedure One player becomes 'question-master' and chooses a word which will be 'taboo'. This player then asks each other player in turn a question, which must be answered without using the forbidden word. The question-master chooses questions which are likely to lead to answers containing the taboo word. The last player left answering questions is the winner, and becomes the next question-master. In an alternative form of the game, the question-master writes the forbidden word on a piece of paper and only reveals it to the other players when someone uses it in an answer. Another alternative is for a taboo *letter* to be chosen, which must not be used in answers.

Example Chas is chosen as the question-master, and he declares that the taboo word is *school*.

CHAS (*to Kate*): Where do you learn things?
KATE: At evening classes.
CHAS (*to Tony*): What did you do when you were a little boy?
TONY: I played with my train set.
CHAS (*to Anna*): What are Eton and Harrow famous for?
ANNA: Their ... er ... academies. (*And so on, until somebody is unwise enough to use the word* school.)

Also called Never Say It.

Tag Wrestling *see* Punchlines.

Talkabout

②③ Two teams of two,
plus a question-master
? Guessing game

✍ Played by speaking

✎ No equipment needed

Object To include chosen words in a short talk, or to guess the subject of the talk from the chosen words.

Procedure The question-master prepares beforehand a list of subjects, each with ten words that might be expected to be used in a short talk on those subjects. For example, if the subject is 'Cats', the expected words might be: *fur, purr, claw, tabby, stroke, milk, bird, tiger, Siamese,* and *dog.* The players are not told what these words are. The question-master sends one team out of the room, and then gives the other team the subject on which they have to speak. Each member of this team has to speak in turn for 20 seconds on the subject, trying to include the words that might be expected in such a talk. As one of the chosen words is mentioned, the question-master crosses it off the list. When the two players have finished their speeches, the question-master calls the other team back into the room and tells them the words that the first team has *not* mentioned. From these words, the second team quickly tries to guess the subject for the talk.

Example The question-master instructs Anna and Tony to leave the room. He then tells Kate and Chas that they have to speak about 'Pencils'. Each player speaks about 'Pencils' for 20 seconds, including the words *wood, write, lead, point,* and *soft,* which the question-master crosses off the list of prepared words. Anna and Tony are called back into the room, and told that the words not used by Kate and Chas are *draw, hard, rub, propelling,* and *graphite.* From these words, Anna immediately guesses that the subject was 'Pencils', winning this round of the game.

Background Popularized on television by Yorkshire TV.

Target *see* Trackword; Words within Words.

Teakettle *see* Coffeepot.

Teapot *see* Coffeepot.

Telegrams

⊘ Any number of players   ✍ Played by writing
? Word-finding game   ✎ Played with pencils and paper

Object To write a short message with words that start with the letters of a chosen word.

Procedure A word is suggested by one of the players, or chosen at random from a book, dictionary, etc. The players then have to write a 'telegram' composed of words that start with successive letters of the chosen word. To add interest to the game, the players may have to write a telegram that is relevant in some way to the chosen word, so that *Christmas* might be turned into: 'COME HOME, RON IMMEDIATELY STOP THE MISTLETOE AWAITS—SAMANTHA.' Alternatively, instead of choosing a word as the basis for the telegram, players can choose a short phrase—or a random series of letters can be suggested by the players in turn.

Example As the first player, Tony chose the phrase *bald head* as the basis for the players to write their telegrams. After five minutes, Kate had written: 'Bitter ageing let Dad's hair escape altogether—Dorothy.' Tony wrote: 'Butter always lets diminishing hair easily almost dazzle.' Anna came up with: 'Baldies—all let's don hairpieces even after dark' and Chas wrote: 'Betty—Alopecia lessens distinction: have ever-decreasing 'air—David.'

Also called Sentence Building; Sentences.

Tennis, Elbow, Foot

⊘ Two or more players   ✍ Played by speaking
? Challenge game; cumulative game   ✎ No equipment needed

Object To think of words associated with previously spoken words.

214

Procedure The players in turn say one word. Each word must be associated in some way with the preceding word. Players can challenge if they do not see the connection with the previous word—and a referee may be chosen to decide if it really is connected or not. Anyone is 'out' who says a word which is unconnected to the preceding word, or hesitates, or repeats a word which has already been used. The winner is the last person left in the game.

Example

TONY: Banana.

ANNA: Yellow.

CHAS: Coward.

KATE: Christmas.

TONY: Challenge!

KATE (*explains*): Christmas is also called 'Noel' and Noel Coward is a
       famous writer. (*The other players accept this and continue.*)

TONY: Mistletoe.

ANNA: Berries.

CHAS: Straw. (*And so on.*)

Compare Associations; Stepping Stones.

The Game *see* Game.

There I Was *see* Punchlines.

Throwing Light

⓿②③ Three or more players	✍ Played by speaking
？ Guessing game	✎ No equipment needed

Object To guess a word that other people are talking about.

Procedure Two of the players choose a word, name, etc. and start a conversation on that subject, without mentioning the word itself. So, if they choose the word *tennis*, one person might say: 'I enjoy it in summer' and the other might reply: 'Yes, but it makes my legs tired.' As they continue to talk, the other players try to guess what they are talking about. Any player who guesses the subject can join in the conversation by making a statement or asking a question which shows that he or she has guessed what it is. The game continues until all the players have joined in the talk.

Titles and Authors *see* Awful Authors.

Tom Swifties

Any number of players	Played by writing or speaking
Punning game; word-finding game	No equipment needed

Object To think of suitable adverbs or verbs to match a statement.

Procedure You can play this game on your own, or a group can take it in turns to think of 'Tom Swifties', which are sentences in which the last word—usually an adverb or verb—sums up a preceding statement, with the help of an atrocious pun. Alternatively, players can set one another sentences which have to be completed with a suitable punning word.

Examples

'I can't find the apples,' said Tom fruitlessly.

'I got the first three wrong,' he said forthrightly.

'Shall I take your picture?' she snapped.

'I'm a burglar,' Tom broke in.

'I deserve to go to prison,' he said with conviction.

When they drained away the water from around my castle, I felt . . . demoted.

Also called Adverbial Puns; Croakers; Pundemonium.

Compare Hermans.

Tongue-twisters

Object To say difficult sentences or phrases clearly several times.

Procedure The players are each given in turn a tongue-twister to pronounce clearly as many times as possible—if desired, within a specified time-limit.

Example Most people know such tongue-twisters as 'She sells sea-shells by the seashore', 'Mixed biscuits', and 'Red lorry, yellow lorry'—and these are all suitable for the game. If these are too simple or too short, you can try longer tongue-twisters, like the one (almost) spoken by Danny Kaye in the film *The Court Jester*: 'The pellet with the poison's in the vessel with the pestle; the chalice from the palace has the brew that is true.'

Tops and Tails

Object To think of words that start and end with particular letters.

Procedure One player asks another player (or all the other players) to think of a word or words that start and end with particular letters. The winner of a round is either the first person to say such a word or the person who writes down the longest list of such words within a given time. Alternatively it can be played as a round game, with each player in turn having to say a suitable word.

Example

ANNA: Can you think of a word that starts with T and ends with L?

TONY: *Trail.*

CHAS: *Tribunal.*

KATE: *Testimonial.* Can you think of a word that starts with Q and ends with T?

ANNA: *Quiet.*

TONY: *Quintet.*

CHAS: *Quest.* Can you give me a word that starts with Z and ends with A?

KATE: *Zebra.*

ANNA: *Zeugma.*

TONY: Er . . . *zippa?*

CHAS: Unacceptable! You are out of this round.

Also called Missing Middles.

Compare Fill-Ins.

Trackword

❶ Any number of players	✍ Played by writing
❓ Grid game; word-finding game	✎ Played with pencils and paper

Object To find words in a grid containing nine letters.

Procedure Players are given a three-by-three grid containing nine letters. Either it is prepared beforehand or the players can make it themselves by choosing a nine-letter word and putting its letters into the grid. The aim of the game is to make as many words as possible, of three or more letters, by moving from one square to the next: going up, down, sideways, or diagonally. You must not go through the same letter more than once in making any one word. If the grid is prepared beforehand for the players, they must also try to find the hidden nine-letter word. In some forms of the game, every word must include the letter in the centre of the grid.

Example

E	I	S
H	P	T
A	T	I

In this trackword, Kate found the words *ape, hat, hip, pat, path, pit, sip, sit, spit, tap, tape, tip,* and *tit.* Anna did even better, finding all these words plus *hep, pith, spa, spat, tapis,* and *the,* as well as the nine-letter word *hepatitis.*

Also called Last Word; Target.

Trailers *see* Follow On.

Trailing Cities *see* Geography.

Transadditions

Any number of players

? Anagrams game; letters game

Played by writing

Played with pencils and paper

Object To create a new word by adding one letter to a word and rearranging the resulting letters.

Procedure This game is similar to Transdeletions, but played in the opposite way. A word is chosen at random, and the challenge is to make a new word by rearranging its letters, with one extra letter added. In the best transadditions, you can continue adding new letters one at a time, and make a new word each time.

Example Take the word *no.* Add the letter E, rearrange the letters, and you can make *one.* Add an S and you can make *nose.* Add C and make *scone.* And so on, through *crones* and *cornets* to *counters.*

Also called Progressive Anagrams; Word-Building.

Compare Isosceles Words; Transdeletions.

Transdeletions

 Any number of players

? Anagrams game;
   letters game

✍ Played by writing

✎ Played with pencils and
   paper

Object To create a new word after removing one letter from a word and rearranging the remaining letters.

Procedure A word is chosen at random. One letter is deleted from it, and the challenge is to make a new word by rearranging the remaining letters. Ideally, letters can continue to be removed from the word one by one, and the remainder turned into a new word each time.

Example Take the word *ridiculous*. If you remove one of the letter I's, you can rearrange the other letters to make the word *ludicrous* (which means the same as *ridiculous*!). Or take the word *retail*. If you remove the letter I, you can rearrange the letters to make the word *alter*. If you take away the letter L, you can shuffle the remaining letters to make *rate*. And so on—through *tea* and *at* to *a*. Even a long word like *emancipator* can be reduced in this way, one letter at a time, until only one letter is left.

Also called Progressive Anagrams.

Compare Add a Letter; Beheadments; Isosceles Words; Shrink Words; Transadditions.

Transformations *see* Doublets.

Transitions *see* Doublets.

Transmutations *see* Doublets.

Traveller's Alphabet

👥 Two or more players
❓ Alphabetical game; word-finding game

✍ Played by speaking
✎ No equipment needed

Object To think of a group of words starting with the same letter of the alphabet.

Procedure The first player starts by saying 'I am going on a journey to A____,' giving the name of a place that starts with A. The next player asks: 'What will you do there?' and the first player has to answer using a verb, adjective, and noun that start with A—such as 'I shall attract anxious ants.' The second player then says 'I am going on a journey to B____,' and the questions and answers continue around the other players in turn. Any player who cannot think of a suitable answering phrase within a reasonable time is out of the game. The winner is the last person left.

Example

ANNA: I am going to Antwerp.
CHAS: What will you do there?

221

ANNA: I shall arrange awful artefacts.

CHAS: I am going to Belize.

KATE: What will you do there?

CHAS: I shall buy beautiful boots.

KATE: I am going to Costa Rica.

CHAS: What will you do there?

KATE: I shall chase cute centipedes.

TONY: I am going to Denmark.

KATE: What will you do there?

TONY: I shall don delightful dungarees. (*And so on.*)

Also called I'm Going to Take a Trip; Travelling Alphabet.

Compare Alphabet Dinner; Grandmother's Trunk; I Packed My Bag; I Went to Market.

Travelling Alphabet *see* Traveller's Alphabet.

Trigrams

⊕²³	Any number of players	✍	Played by writing
?	Word-building game	✎	Played with pencils and paper

Object To think of words containing three given letters.

Procedure One player calls out three letters of the alphabet, and all the players have to write down (in a set time) all the words they can think of that contain those three letters together. The winner is the player who thinks of the largest number of words. Alternatively, players score one point for every word they make but two points for words that nobody else has written down. If desired, it can be ruled that the letters must not start or end words.

Example Kate calls out the letters R–I–N. After five minutes, Tony has written down: *bring, string, cringe, meringue, marine, brine, herring, prince, ring,* and *earring*. However, Anna wins this round because she has written down all these words as well as *fringe, gringo, grin, crinkle, wrinkly, principality, syringe, trinity, trinket, urine, wring, crinoline, terrine, raring, perinatal,* and *rinderpest*.

Compare Digrams.

Triple Acrostics

👥 Any number of players
❓ Word-finding game

✍ Played by writing
✎ Played with a previously prepared puzzle, plus pencils and paper

Object To solve clues leading to words whose first, last, and middle letters spell out words.

Procedure Triple acrostics are similar to *Double Acrostics* but, in a triple acrostic, three words or phrases are spelt out by: (1) the first letters of the mystery words, (2) the last letters of these words, and (3) a series of letters in the middle of these words. Clues are given to the words, and the clues are sometimes given in rhyme, as in the example below.

Example

Left, middle, and right
Give us choice of a light.

(1) The kind of glance which he who's lost his heart
Bestows on her who wears the latter part.

(2) Here is one
With a gun.

(3) This is bound
To go round.

(4) Simplify taste
And eliminate waste.

(5) My meaning is made plain
By my saying it again.

Solution

A   d   o   R   i   n   G

M   u   s   k   E   t   e   e   R

B   a   n   D   a   g   E

E   c   o   n   O   m   i   z   E

R   e   i   t   e   R   a   t   i   o   N

Triple Meanings

⚉ Two or more players   ✍ Played by speaking
? Guessing game   ✎ No equipment needed

Object To guess words which have three different meanings.

Procedure One player thinks of a word which has three different meanings, such as *boom*—which can mean a loud noise, a long pole, and a period of sudden prosperity. This player then asks the other player or players to guess the word from these definitions. The player who guesses correctly chooses the next mystery word.

Example

ANNA: Which word has these three meanings: a bird's beak, an invoice, and a kind of poster?

TONY: *Bill.* Which word means all these three things: your sweetheart, great affection, and a score of nothing?

CHAS: *Love.* Can you guess this word? It means enthusiastic, very cold, and a funeral song.

KATE: *Keen.*

Compare Homonyms.

Triplets

⚉ Any number of players   ✍ Played by writing
? Challenge game   ✎ Played with pencils and paper

Object To change one word into another in stages, each stage making a new word.

Procedure This game is similar to Doublets, except that the two words can be of different lengths, and players change the number of letters as they move from one word to another. Players are given, or choose, one word to change into another. The two words should preferably be related in sense: like *fast* and *slow*, or *dustpan* and *brush*. Players add or subtract one letter at a time and, at each stage, a new word has to be made. No word may be repeated. The winner is the person who completes the change in the smallest number of moves.

Example Change *first* to *last*:

FIRST
FIST
FIT
IT
AIT
AT
EAT
EAST
LEAST
LAST

Compare Doublets.

Background Invented by Peter Newby.

Trumps *see* Bullets.

Twenty Questions

Two or more players (or two teams)	Played by speaking
Guessing game	No equipment needed

Object To guess a word by asking no more than 20 questions.

Procedure One player thinks of a word: it can be an object, a living being, or something abstract. This player tells the others if it is animal, vegetable, mineral, or abstract. 'Animal' means an animal, a person, or something derived from an animal (e.g. a lucky rabbit's foot). 'Vegetable' covers plants and things that originate from growing matter (e.g. a swede, a potato crisp, or a book). 'Mineral' refers to inorganic objects and substances (e.g. a television set). 'Abstract' covers qualities (such as bravery) and such things as the titles of films, songs, etc. Things can be combinations of more than one of these categories (e.g. a bag of crisps would be 'vegetable and mineral').

The players then ask the first player questions to try to discover what the thing is. The answers must be confined to *yes* or *no* (*I don't know* is sometimes allowed). If someone guesses the word when no more than 20 questions have been asked, he or she chooses the next word for guessing. If nobody guesses the word, the first player chooses another one. If the game is played by two teams, they can alternate in choosing and guessing the word.

Example Chas starts, and says that he is thinking of something which is 'vegetable'.

TONY: Is it a flower?
CHAS: No.
KATE: Does it grow in the garden?
CHAS: No.
ANNA: Is it a house plant?
CHAS: No.
KATE: Can you eat it?
CHAS: Yes.
TONY: Is it a fruit?
CHAS: No.
ANNA: Can you cook it?
CHAS: Yes.
TONY: Is it a pudding?
CHAS: No.
KATE: Would you eat it as the main dish of a meal?
CHAS: Yes.
ANNA: Is it an omelette?
CHAS: No.
KATE: Is it a nut cutlet?
CHAS: Yes!

Also called Animal, Vegetable, and Mineral; Yes and No; Yes or No.

Compare Who Am I?

U

Ultraghost *see* Catchword.

Uncrash

❶ ②③ Two or more players	✍ Played by writing
❓ Challenge game; cumulative game	✎ Played with pencils and paper

Object To think of words that do not share any letters with words earlier in a series.

Procedure This game is like a reversal of the game *Crash*. One player writes down a short word—of, say, three or four letters. The next player has to write beneath it another word of the same length which uses none of the same letters in the same position as they occur in the first word (so, if the first word is *pit*, you can write *top* but not *sin*). The players continue in turn adding words which include none of the letters that have already been used in that part of the word. The winner is the last player able to add a word to the list (alternatively, the loser is the first player who cannot add a word to the list).

Example Anna writes down the word *luck*. Beneath it Kate writes *boss*. The other players each add a word, building up a column that looks like this:

L U C K

B O S S

D A W N

R I P E

S P I T

P R O D

M E A L

T H E M

C L U B   Kate could not think of a word to add beneath *club*.

Univocalics

⊕②③ Any number of players   ✍ Played by writing
? Challenge game   ✎ Played with pencils and paper

Object To write something using only one vowel.

Procedure Univocalics are pieces of writing which deliberately use only one vowel of the alphabet. If you find it easy to do this in prose, try writing a poem using only one of the vowels.

Example Anna always asks all fans to call any day at half past ... (the writer finds it difficult to continue this piece, as there are no numbers under a thousand which use the letter A).

Compare Lipograms.

UnScrabble *see* Scrabble.

V

Verbal Sprouts *see* Arrow of Letters.

Vicar's Cat *see* Minister's Cat.

Vocabularyclept Poetry

⊕ Any number of players	✍ Played by writing
? Challenge game	✎ Played with pencils and paper

Object To reconstruct a poem after its words have been mixed up.

Procedure A poem (preferably a fairly well-known poem) is chosen and its words are mixed up: either entirely at random or rearranged in alphabetical order. The players then try to reconstruct the poem, or make a new poem from the words.

Example These are the words of one stanza of a poem, rearranged in alphabetical order: *a, age, along, and, dew, dim, drenched, evening, goes, in, nod, of, old, road, rose, shepherd, softly, the, the, twilight, with, with, with, wrinkled.* How can they be put back in order?

Solution

> Softly along the road of evening,
> In a twilight dim with rose,
> Wrinkled with age, and drenched with dew,
> Old Nod, the shepherd, goes.
>
> *Nod*, by Walter de la Mare

Also called Anapoems; Constructapo.

Vowel Play *see* Missing Vowels.

W

Wellerisms

⊕ Any number of players

? Punning game

✍ Played by writing or speaking

✎ Played with pencils and paper, or with no equipment

Object To think of humorous explanations of how phrases came to be spoken.

Procedure Wellerisms are named after a character in Charles Dickens's *Pickwick Papers* called Sam Weller, who said things like: 'Out with it, as the father said to the child when he swallowed a farthing.' Players suggest sentences or phrases for other players to complete with an explanation starting with 'as. . .'. The sentences or phrases are usually idioms, proverbs, or clichés, and the completions often use puns. The game can be played by each player writing a sentence or a phrase on a piece of paper, and passing it to the next player to complete.

Example Tony wrote down the sentence: 'It won't be long now,' which Anna completed by writing: 'as the monkey said when it cut off its tail'. Anna wrote the sentence: 'My lips are sealed,' which Chas completed by writing: 'as the man said when he swallowed the glue' (but Kate suggested the alternative: 'as the ringmaster said when he was kissed by the performing seal').

What Did We Say? *see* Shouting Proverbs.

What is My Thought Like?

👥 Two or more players	✍ Played by writing and speaking
❓ Guessing game	✎ Played with pencils and paper

Object To explain how one's guess at a word is connected with the actual word.

Procedure The first player secretly writes down a word for a thing or the name of a person. He or she then asks the other players: 'What is my thought like?' and notes down their answers. The first player then reveals the secret 'thought' and the other players in turn have to explain how their guess resembles the 'thought'.

Example Kate asks the other players: 'What is my thought like?' Anna replies *a light*; Chas suggests *a fruit cake*; and Tony says *a kitten*. Kate then reveals that her thought was 'Tony' and she asks Anna: 'Why is Tony like a light?' Anna replies: 'Because he often goes out.' Kate asks Chas: 'Why is Tony like a fruit cake?' and Chas answers: 'Because he's nutty.' Kate asks Tony: 'Why is Tony like a kitten?' and Tony replies: 'Because he's so lovable!'

Also called Analogies.

What is the Question?

👥 Two or more players	✍ Played by writing or speaking
❓ Challenge game	✎ Played with pencils and paper, or with no equipment

Object To think of the question to match a particular reply.

Procedure Players give each other answers to questions. The questions are not revealed: it is the job of the other players to think of a suitable question to match each given answer. The answer will usually be a familiar phrase, a celebrity's name, a book title, etc. The expected questions often contain puns.

Example Chas gives to Tony the answer 'Beethoven's fifth', expecting Tony to come up with some such question as 'How many children has Ludwig got now?' Tony supplies an equally acceptable question: 'Where is Ludwig in the race?' Tony then gives Chas the answer 'The cradle will rock'. Chas supplies the question: 'What do you say to William Rock when he asks you where to put the baby at bedtime?'

Where, When, and How *see* How, When, and Where.

Who Am I?

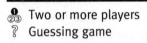 Two or more players	✍ Played by speaking
? Guessing game	✎ Played with pencils and paper, or with no equipment

Object To guess a hidden identity.

Procedure This game is played in several different ways. One form is described in the entry for *Botticelli*. A simpler form is for all the players except one to agree on a famous character: the remaining player then tries to identify that character by asking questions of the other players in turn. The questions must be answered simply by *yes* or *no*. Usually the player must guess the identity before 20 questions have been asked. Alternatively, each player can give the guesser one clue about the chosen character. Another method is for one player to adopt a personality and for all the other players to try to guess who it is.

In yet another way of playing this game, the names of different characters are written on pieces of paper which are pinned on the backs of all the players. The players then try to guess 'who they are' by asking the other players questions about their identity. Again, the questions must be answered either *yes* or *no*. Any player who guesses the identity can remove the card from his or her back. The game can be played in pairs if preferred.

Example Tony goes out of the room, while the other players choose a secret personality. When Tony returns, he asks them questions in turn:

TONY: Am I living?
ANNA: No.
TONY: Am I famous?
KATE: Yes.
TONY: Am I female?
CHAS: No.
TONY: Am I involved in the arts?
ANNA: Yes.
TONY: A composer?
KATE: No.
TONY: A painter?
CHAS: No.

(And so on. Tony uses up 20 questions but surprisingly fails to identify William Shakespeare.)

Also called Who Are You?

Compare Botticelli; Twenty Questions.

Who Are You? *see* Who Am I?

Who Wrote What *see* Awful Authors.

Why Do You Like It? *see* How, When, and Where.

Wild Crash

 Two players
? Guessing game

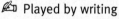 Played by writing
✎ Played with pencils and paper

Object To guess a five-letter word chosen by one's opponent.

Procedure The game is the same as *Crash*, except that a player can keep changing the hidden word, so long as it fits the facts that have been given to the other player about its letters.

Win, Lose, or Draw *see* Pictures.

Winners *see* Bullets.

Word Alchemy *see* Doublets.

Word Associations *see* Associations.

Word Battleships

② Two players	✍ Played by writing
❓ Grid game; guessing game	✎ Played with pencils and paper

Object To guess words hidden in a grid.

Procedure Both players draw a grid—say, six squares by six. They number the columns of the grid from one to six along the top, and from A to F down the side, so that each square can be identified by a number and letter (e.g. the square in the top left-hand corner is 1A; the square in the bottom right-hand corner is 6F). Each player then writes an agreed number of words somewhere in their grid—they might choose one word each of six, five, four, and three letters. The words can interlock, as in a crossword. If agreed, the words can all have a common theme, or they can compose a well-known proverb or saying. The remaining squares are left blank. The players also write on their pieces of paper a second, blank grid of six by six, in which to write their opponent's letters as they discover them.

Each player in turn calls out the number and letter of a square, and their opponent says what letter is in that square - or if it is blank. In this way each player can build up a picture of what letters their opponent's grid contains. Instead of choosing a square, a player can guess the opponent's hidden word: if it is guessed correctly, the guesser enters the whole word in the grid; if not, the game continues. The winner is the player who guesses all the opponent's words first.

An alternative method of play is for players to call out a letter of the alphabet, in which case all occurrences of that letter are entered in the grid. If the game is played this way, it is best for the players to keep a note of the letters they have already called out.

Example Kate and Chas agree to use grids of six squares by six, and to have the theme of 'animals'. Chas draws his grid like this:

	1	2	3	4	5	6
A			D			
B			O			O
C			N	E	W	T
D			K			T
E	A	P	E			E
F			Y			R

Kate starts by calling out 3D, choosing a square near the centre of the grid, hoping that some letters are likely to be near the centre. This gives her the K, but she then wastes her next two turns by calling out 2D and 4D, wrongly imagining that the K is part of an 'across' word. Realizing that the K is part of a down word, Kate next calls out 3C, which gives her NK. Since this is unlikely to be the start of an animal's name, she next chooses 3B, which gives her ONK. She wrongly guesses that 3A must be an M (assuming the word is *monkey*) but then rightly guesses that the word is *donkey*. And so on.

Also called (as a boxed game) Battle Words (Copyright William Maclean's Games 1993).

Compare Get the Message; Quizl.

Word Builder *see* Words within Words.

Word-Building *see* Transadditions.

Word Chain *see* Follow On.

Word Chains *see* Doublets.

Word Diamonds *see* Word Squares (1).

Word Divisions

② Any number of players

? Wordplay game

✍ Played by writing or speaking

✎ Played with pencils and paper, with newspapers and magazines, or with no equipment

Object To find or devise strange word divisions.

Procedure Several forms of wordplay are possible with word divisions, especially those that occur in print when a word is divided between two lines. One can look in newspapers and magazines for ridiculous or misleading word divisions, such as *rat-her, ado-ration, hat-red, since-rely, mole-station, leg-end, mans-laughter, Some-rset*, and *Chic-ago*.

Players can also try to make up their own weird divisions; try to think of words that sound like two other words when divided (for example, *gruesome* which sounds like *grew some*); or devise phrases like *Was Ted wasted?* or *sentry's entry* which repeat the same words divided differently.

In another game, one player makes up a sentence for other players to solve. The sentence will contain blanks that can be filled by a word or phrase divided in two or more different ways. Thus in 'I am – – – – – – – – that Henry VIII could never be described as a —— ——,' the blanks can be filled by *thinking* and *thin king*.

Compare Fusions.

Word Endings *see* Endings.

Word for Word

② Two teams of players

? Challenge game

✍ Played by speaking

✎ No equipment needed

Object To think of a series of words which are unconnected in meaning.

Procedure The players divide into two teams, preferably with one person acting as referee. The first member of the first team says a word, and the second member has to follow with a word which is entirely unconnected in meaning. Members of the other team can challenge if they think there *is* a connection: if the judge upholds their challenge, the play passes to the other team.

Example Anna and Tony form one team; Kate and Chas form the other.

ANNA: Brick.
TONY: Sausage.
ANNA: Sock.
TONY: Wind.
KATE: Challenge! A wind-sock is used at airports to show the wind direction.
TONY: You are right. Now it's your team's turn.
KATE: Custard.
CHAS: Plinth.
KATE: Serendipity.
CHAS: Spatula. (*And so on.*)

Compare Associations.

Word Golf *see* Doublets.

Word Hunt *see* Words within Words.

Word Ladders *see* Doublets.

Word Links *see* Doublets.

Word-Making (1)

❶ Two or more players	🖎 Played on a table or floor
❓ Letters game; word-building game	🖎 Played with a set of letters on cards or tiles

Object To make words from letters which are shown one-by-one.

Procedure The tiles from Scrabble or the cards from Lexicon can be used for this game. Alternatively, someone can make a set of cards out of cardboard, each containing one letter of the alphabet. The set should include lots of vowels and other common letters like R and S. The cards or tiles are dealt out equally to the players, who have them face-down in front of them. One by one, the players turn up one of the letters in front of them, placing it in the middle of the table or floor. If anyone can make a word from the revealed letters that accumulate, he or she takes those letters and spells out the word. The winner is the player who has made the most words when the cards or tiles run out, or the player whose words contain the highest total of letters.

Word-Making and Word-Taking is played in the same way, except that players can take other people's words if they can use them in making longer words.

Example The cards are dealt out to the players, who agree that words must contain at least four letters each. Tony starts by turning up the letter P; Anna turns up E; Kate turns up L; Chas turns up A, and shouts out the word *plea* which he sees they can make. He takes the four letters and lays out the word *plea* in front of him. Tony turns up S and Anna shouts out *sepal,* which she sees can be made from the S and the letters in *plea.* She lays out *sepal* in front of herself. Anna turns up J but nobody can see any way of using it. Kate turns up C, shouts out *places,* and takes the letters of *sepal* and lays out *places* in front of her. And so on . . .

Word-Making (2) *see* Words within Words.

Word Master Mind *see* Jotto.

Wordnim

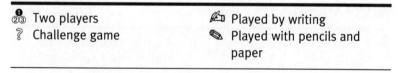

👥 Two players		✍ Played by writing	
❓ Challenge game		✎ Played with pencils and paper	

Object To use up letters of the alphabet by making words.

Procedure This is a verbal form of the game Nim, in which players remove objects from a pile and try to force their opponent to take the last object. Wordnim can be played in several different ways:

1. The players write down a list of all the consonants in the alphabet (i.e. every letter except for A, E, I, O, and U). They then take turns to write down single words which include particular letters of the alphabet, which are crossed off the alphabet list. Every word must only include vowels plus any of the consonants that have not yet been crossed out. For example, the first player might write down *spectacle* and cross out the C, L, P, S, and T. The second player might write down *grade* and cross out the D, G, and R. The first player could then write *money* and cross out the M, N, and Y. The second player might write *behave* and cross out the B, H, and V. The first player is now left with the consonants F, J, K, Q, W, X, and Z, and decides to write down *few*. This leaves the second player to make a word using the consonants J, K, Q, X, and Z. He cannot think of one, so he loses the game - but he could have thought of *quiz* and saved the game.

2. The players write down the whole alphabet (including the vowels). They then take turns to write down words, beginning with a word that starts with A, which is crossed out. If the word also includes B, that letter is crossed out. If it also includes C following the B, the C is crossed out—and so on with subsequent letters. So, if the first player writes down *abacus*, the A, B, and C are deleted; if he writes down *abscond*, the A, B, C, and D are deleted. The second player then writes down a word starting with the next letter that has not yet been deleted. So, if the first player wrote *abscond*, the second player might write *deflated* and cross out the D, E, and F. The loser is the player who has to write down the last word, which will usually start with Z.

3. The game is played as under '2' above but the first player can start with *any* letter in the alphabet. The players then have to work through the alphabet until they return to the starting-point. Players can decide beforehand if the person who writes down the last word is to be considered the winner or the loser.

4. The game can be played in the same way, but the winner can be the person who crosses out the largest number of letters.

239

5. The players choose a number of words—say two each—and they take turns to remove from the group of words any set of letters that can be used as an anagram of a word. For example, if the words are *table*, *chair*, *bottom*, and *apple*, a player might cross out the letters T, C, H, I, and P to make the word *pitch*. The loser is the player who cannot make a word out of the remaining letters when it is his or her turn. The game can be made harder by forcing players to take letters from only one of the words—and even harder by insisting that you must be able to make a word out of the letters of a word which are *not* deleted.

Also called Nymphabet.

Word Order   *see* Associations.

Word Parade

 Five or more players
? Active game;
   challenge game

🖾 Played with cards
✎ Played with letters on cards

Object   To spell out a word in teams.

Procedure   One player is chosen as the leader, and the remaining players divide into two teams. The leader writes letters of the alphabet on pieces of card (e.g. postcards). There should be two duplicate sets of letters—one set for each team—with either one or two cards for each player. For example, if there are five players in each team, the leader can prepare two sets of ten cards, bearing ten different but fairly common letters of the alphabet The letters in each set are dealt out at random to the players in each team, so that each player is holding two cards. The leader then shouts out a word which can be made from the letters in the sets, and the players holding the letters in that word have to hold up their cards so as to spell out the word. The first team to spell the word correctly (without any extra letters) wins one point. The game continues in this way for an agreed number of rounds.

Word Ping-Pong

🎱 Two or more players   ✍ Played by writing
❓ Cumulative game   ✎ Played with pencils and paper

Object To change one word into another, one letter at a time.

Procedure Word Ping-Pong can simply be the same as Doublets. It is also the name for a game using the principle of Doublets to make a contest between two players. One player (the 'server') writes down a four-letter word. The second player has to change either the third or fourth letter to make a new word. The first player then has to change either the first or second letter of the new word to make a third word, and so on, until one player is unable to create a new word. The other player then wins one point. Service changes every five points, and the winner is the first player to reach 21 points. If the score reaches 20–20, the service alternates between the players until one of them wins a clear lead.

The first word written must be capable of being changed into another word by a change in its last two letters. The first player can only change one of the first two letters; the second player can only change one of the last two letters. Neither player may introduce the same letter more than three times in a particular position (i.e. first, second, third, or fourth letter), and the same word cannot be used more than once. The game can be played by more than two people, either in two teams or with each player in turn changing a letter.

Example Anna starts by writing down the word FOUR. Kate can change either the U or the R, so she changes the R to L, making FOUL. Anna can then change the F or the O: she changes the F to S, making SOUL. The diagram on the next page shows how the ladder builds up.

```
F    O    U    R
F    O    U    L
S    O    U    L
S    O    I    L
S    A    I    L
S    A    I    D
L    A    I    D
L    A    N    D
H    A    N    D
H    A    N    G
H    U    N    G
H    U    N    T
H    I    N    T
H    I    L    T
W    I    L    T
```

The game continues, but it becomes increasingly difficult for both players, as neither of them can use the same letter more than three times, which gradually reduces the possible words they can make.

Word Pyramids *see* Word Squares (1).

Words *see* Crash.

Word Search

 Any number of players
? Grid game;
  word-finding game

✍ Played by writing
✎ Played with a grid in which
  words are hidden

Object To find words concealed in a grid.

Procedure This game is usually played with prearranged grids printed in word-puzzle magazines, but players can construct them for other players to solve. The grid contains a jumble of letters. Players have to work their way through the grid—horizontally, vertically, or diagonally, backwards or forwards—searching for the hidden words. For instance, in the grid below, the word *cricket* is spelt vertically down the right-hand column. The hidden words are usually related in sense or subject: e.g. by being words concerned with a particular area of interest. Sometimes all the hidden words are listed next to the grid, to make them easier to find. When you find a word, mark it on the grid by drawing a line through all its letters.

Example Try to find the 20 listed sports and games hidden in the grid:

bingo	monopoly
brag	poker
crib	polo
cricket	pontoon
golf	pool
hoopla	rummy
jotto	snap
loo	snooker
lotto	tag
ludo	yoga

O	G	N	I	B	B	I	R	C
D	Y	O	G	A	R	D	U	R
U	H	P	L	Z	E	A	S	I
L	O	O	C	F	K	B	G	C
M	O	N	O	P	O	L	Y	K
P	P	T	X	T	O	L	L	E
A	L	O	T	E	N	K	O	T
N	A	O	A	O	S	J	E	P
S	J	N	G	Y	M	M	U	R

Word Squares (1)

⊕⊕ Any number of players   ✍ Played by writing
? Grid game   ✎ Played with pencils and paper

Object To build up words in letters that make a square.

Procedure Word squares consist of squares formed by words which read the same horizontally or vertically, like this:

B   A   R   E

A   M   E   N

R   E   A   D

E   N   D   S

Sometimes, word squares have *different* words across and down, as in this example:

B   A   S   H

R   U   L   Y

I   T   E   M

G   O   W   N

There are several ways of using word squares in games. You can try constructing your own, starting with fairly easy squares of four-letter words and working up to the harder problem of constructing a satisfactory square of seven-letter words (or even longer). An extra challenge is to devise a four-by-four or five-by-five square which uses no letter more than once (as is the case with the second example above). Or you can attempt to solve word squares created by other people, in which case they may supply clues to the words in the square.

Puzzles similar to word squares can be made in other shapes, such as diamonds, pyramids, and stars.

Word Squares (2)

Object To build up words in two five-by-five grids.

Procedure Each player draws a grid of 25 squares, five across and five down. The first player calls out a letter, which both players enter anywhere in their own grids. The players call out letters alternately and try to build up words, across or down. When each grid is full, five points are scored for five-letter words, four points for four-letter words, and so on. The player with the higher total of points is the winner. To make the game more complex, a grid of more than five-by-five squares can be used.

Example Tony starts by calling out Q, which he thinks will make the game difficult for his opponent. But Kate knows that many words start with QU–, so she calls out U. Tony chooses I, intending to make *quite* but he is puzzled when Kate says R, not realizing that she can make *quire*. So Tony starts another line with the R and calls out V, which he intends eventually to build up into *raven*. Kate inserts the V under the U, intending to build up *uvula*. One can guess who eventually won the game.

Compare Crossword Game; Ragaman.

Word Stars *see* Word Squares (1).

Word Substitution

Object To replace the nouns in a piece of writing with other nouns later in a dictionary.

Procedure In this game, players see what happens when you take a piece of writing and replace every noun with the seventh noun following it in a chosen dictionary. The result is sometimes nonsensical but it can also be surprisingly meaningful.

Example A player might choose at random the following verses by Wordsworth:

> I wandered lonely as a cloud
> That floats on high o'er vales and hills.
> When all at once I saw a crowd,
> A host, of golden daffodils.

By replacing each noun with the seventh noun after it in a dictionary, the player might create this new verse:

> I wandered lonely as a clown
> That floats on high o'er valentines and hims.
> When all at once I saw a crucifix,
> A hostelry, of golden dahlias.

Also called N plus Seven; S plus Seven.

Compare Semantic Poetry.

Background A pastime devised by members of OuLiPo, a French group interested in wordplay.

Words within Words

👥 Any number of players	✍ Played by writing
❓ Word-finding game	✎ Played with pencils and paper

Object To make words from the letters in a word.

Procedure A word is chosen—usually a fairly long word (*Constantinople* is an old favourite)—and players have to make as many words as they can from the letters in the chosen word. The letters can be used in any order, but a letter can be used in any one word only as many times as it occurs in the chosen word (so from the chosen word *teacher* you can make *reheat* but not *treat*). A time-limit can be set (say, five minutes), and the winner is either the person who finds the largest number of words, or the person who finds the largest number of words that other players have *not* found.

Example The players open a book and pick out a word at random. It is *references*. Tony makes the following words from its letters: *refer, fence, fencer, fen, free, ref, serf, err, erne, see, seer, scene,* and *referee.* Kate wins by finding more words than these; her list contains: *refer, fern, free, fee, reef, reefer, referee, screen, scree, scene, see, seen, seer, sneer, sere,* and *serene*. She could have scored even higher if she had noticed such words as *censer, cere, creese, feces, nee,* and *ne'er.* She might also have noticed *Cree,* but proper names are not usually allowed in this game.

Also called Constantinople; Dictionary; Hidden Words; In-Words; Keyword; Key Words; Making Words; Multiwords; Target; Word Builder; Word Hunt; Word-Making; Wordy Word.

Words Worth *see* Jotto.

Wordsworth *see* Crossword Game.

Wordy Word *see* Words within Words.

Wraiths *see* Ghosts.

Wuthering Hillocks *see* Deflation.

Y

Yes and No (1)

👥 Two or more players   ✍ Played by speaking
❓ Challenge game   ✎ No equipment needed

Object To answer questions for one minute without saying 'yes' or 'no'.

Procedure One player asks another player a series of quick questions, which that player has to answer without using the words *yes* or *no* or nodding the head. Either one player can assume the role of 'question-master' and ask each player questions in turn, or the players can quiz each other in turn. The winner is the player who lasts longest without saying the forbidden words. In a variation of the game, the words *you* and *I* also have to be avoided.

Another way of playing the game is for each player to be given five matchsticks, counters, or other tokens. The players act in turn as question-master. Each time anyone is tricked into saying *yes* or *no*, that player is given one token by the questioner. The winner is the first player to get rid of all his or her tokens. If desired, every player can move on to a new partner each time they give or take a token.

Example

CHAS: What is your name?
TONY: Tony.
CHAS: Are you sure?
TONY: Of course I'm sure.
CHAS: Where do you live?
TONY: Oxford.
CHAS: So you live in Oxford, do you?
TONY: I do.
CHAS: Is Oxford a nice place?
TONY: It is.
CHAS: You like it, do you?
TONY: Yes, I do ...
CHAS: You said 'yes'!

Yes and No (2) *see* Twenty Questions.

Yes–No *see* Yes and No (1).

Yes or No *see* Twenty Questions.

Yessir, Nossir *see* Yes and No (1).